Christian Bonatto Minella

Transition Metal Fluorides on Ca(BH4)2-systems

Christian Bonatto Minella

Transition Metal Fluorides on Ca(BH4)2-systems

Südwestdeutscher Verlag für Hochschulschriften

Impressum / Imprint
Bibliografische Information der Deutschen Nationalbibliothek: Die Deutsche Nationalbibliothek verzeichnet diese Publikation in der Deutschen Nationalbibliografie; detaillierte bibliografische Daten sind im Internet über http://dnb.d-nb.de abrufbar.
Alle in diesem Buch genannten Marken und Produktnamen unterliegen warenzeichen-, marken- oder patentrechtlichem Schutz bzw. sind Warenzeichen oder eingetragene Warenzeichen der jeweiligen Inhaber. Die Wiedergabe von Marken, Produktnamen, Gebrauchsnamen, Handelsnamen, Warenbezeichnungen u.s.w. in diesem Werk berechtigt auch ohne besondere Kennzeichnung nicht zu der Annahme, dass solche Namen im Sinne der Warenzeichen- und Markenschutzgesetzgebung als frei zu betrachten wären und daher von jedermann benutzt werden dürften.

Bibliographic information published by the Deutsche Nationalbibliothek: The Deutsche Nationalbibliothek lists this publication in the Deutsche Nationalbibliografie; detailed bibliographic data are available in the Internet at http://dnb.d-nb.de.
Any brand names and product names mentioned in this book are subject to trademark, brand or patent protection and are trademarks or registered trademarks of their respective holders. The use of brand names, product names, common names, trade names, product descriptions etc. even without a particular marking in this works is in no way to be construed to mean that such names may be regarded as unrestricted in respect of trademark and brand protection legislation and could thus be used by anyone.

Coverbild / Cover image: www.ingimage.com

Verlag / Publisher:
Südwestdeutscher Verlag für Hochschulschriften
ist ein Imprint der / is a trademark of
OmniScriptum GmbH & Co. KG
Heinrich-Böcking-Str. 6-8, 66121 Saarbrücken, Deutschland / Germany
Email: info@svh-verlag.de

Herstellung: siehe letzte Seite /
Printed at: see last page
ISBN: 978-3-8381-2792-7

Zugl. / Approved by: Hamburg-Harburg, TU, Diss., 2013

Copyright © 2014 OmniScriptum GmbH & Co. KG
Alle Rechte vorbehalten. / All rights reserved. Saarbrücken 2014

Contents

1 Introduction   **1**
    1.1 $Ca(BH_4)_2$ 13
    1.2 $Ca(BH_4)_2 + MgH_2$ 15
    1.3   Aim of the work 16

2 Experimental Part   **19**
    2.1   Materials 19
       2.1.1   Chemical Synthesis of $CaB_{12}H_{12}$ 20
    2.2   Sample preparation 20
       2.2.1 $Ca(BH_4)_2$ with additives 20
       2.2.2 $Ca(BH_4)_2 + MgH_2$ with and without additives 20
    2.3   Kinetic characterisation 21
    2.4   Thermal analysis 22
    2.5   Infrared Spectroscopy 23
    2.6   *Ex-situ* X-ray Diffraction 23
    2.7   *In-situ* Synchrotron Radiation Powder X-ray Diffraction 24
    2.8   X-ray absorption spectroscopy 25
    2.9 $^{11}B\{^1H\}$ Magic Angle Spinning-Nuclear Magnetic Resonance . . . 26
    2.10   Transmission Electron Microscopy 26

3 Results   **27**
    3.1 $Ca(BH_4)_2$ 27
       3.1.1   The (De)hydrogenation Reaction 27
       3.1.2   Thermal Analysis-Mass Spectrometry 30
       3.1.3   *In-situ* Synchrotron Radiation Powder X-ray Diffraction . . 31
       3.1.4   The (Re)hydrogenation Reaction 33
       3.1.5 $^{11}B\{^1H\}$ Magic Angle Spinning-Nuclear Magnetic Resonance . . 34
       3.1.6   Transmission Electron Microscopy 35

3.2 Effect of Transition metal fluorides on the sorption properties of $Ca(BH_4)_2$ . . . 37
    3.2.1 The (De)hydrogenation Reaction 39
    3.2.2 Thermal-Analysis 43
    3.2.3 The (Re)hydrogenation Reaction 45
    3.2.4 $Ca(BH_4)_2$ + NbF_5: *in-situ* Synchrotron Radiation Powder X-ray Diffraction 48
    3.2.5 $Ca(BH_4)_2$ + TiF_4: *in-situ* Synchrotron Radiation Powder X-ray Diffraction 50
    3.2.6 $Ca(BH_4)_2$ + NbF_5: X-ray Absorption Near Edge Structure 52
    3.2.7 $Ca(BH_4)_2$ + TiF_4: X-ray Absorption Near Edge Structure. 54
    3.2.8 ^{11}B {^1H} Magic Angle Spinning-Nuclear Magnetic Resonance. 55
    3.2.9 Transmission Electron Microscopy 57
3.3 Effect of the Ti-isopropoxide on the sorption properties of $Ca(BH_4)_2$ 60
    3.3.1 The (De)hydrogenation Reaction 61
    3.3.2 Thermal-Analysis 63
    3.3.3 The (Re)hydrogenation Reaction. 65
    3.3.4 ^{11}B {^1H} Magic Angle Spinning-Nuclear Magnetic Resonance. 68
    3.3.5 Transmission Electron Microscopy 69
3.4 Effect of the CaF_2 on the sorption properties of $Ca(BH_4)_2$ 71
    3.4.1 The (De)hydrogenation Reaction. 71
    3.4.2 Thermal-Analysis. 73
    3.4.3 The (Re)hydrogenation Reaction 74
3.5 $Ca(BH_4)_2$ + MgH_2 76
    3.5.1 The (De)hydrogenation Reaction 76
    3.5.2 Thermal Analysis. 78
    3.5.3 *In-situ* Synchrotron Radiation Powder X-ray Diffraction 79
    3.5.4 The (Re)hydrogenation Reaction. 84
    3.5.5 ^{11}B {^1H} Magic Angle Spinning-Nuclear Magnetic Resonance . . . 87

3.5.6 Is the formation of $Ca_4Mg_3H_{14}$ phase a necessary reaction step during the decomposition of a $Ca(BH_4)_2 + MgH_2$ composite? 90

3.5.7 Is the hydrogen back pressure influencing the decomposition path of the $Ca(BH_4)_2 + MgH_2$ composite system? 93

3.6 Effect of NbF_5 and TiF_4 on the sorption properties of the $Ca(BH_4)_2 + MgH_2$ composite system 98

3.6.1 The (De)hydrogenation Reaction. 98

3.6.2 Thermal-Analysis. 101

3.6.3 The (Re)hydrogenation Reaction. 102

3.6.4 $Ca(BH_4)_2 + MgH_2 + NbF_5$: *in-situ* Synchrotron Radiation Powder X-ray Diffraction 103

3.6.5 $Ca(BH_4)_2 + MgH_2 + TiF_4$: *in-situ* Synchrotron Radiation Powder X-ray Diffraction 105

3.6.6 $Ca(BH_4)_2 + MgH_2 + NbF_5$: X-ray Absorption Near Edge Structure . 106

3.6.7 $Ca(BH_4)_2 + MgH_2 + TiF_4$: X-ray Absorption Near Edge Structure . . 108

3.6.8 ^{11}B {^1H} Magic Angle Spinning-Nuclear Magnetic Resonance. . . 110

4 Discussion   113

4.1 $Ca(BH_4)_2$ system 113

4.2 Role of additives on the $Ca(BH_4)_2$ system 116

4.3 $Ca(BH_4)_2 + MgH_2$ system 121

4.3.1 Reaction Scheme for the pure milled $Ca(BH_4)_2 + MgH_2$ composite system 124

4.4 Transition metal fluorides doped $Ca(BH_4)_2 + MgH_2$ composite system . . 125

5 Summary and Outlook   127

6 Bibliography   131

7 Acknowledgments   141

1 Introduction

After the discovery of the steam engine, at the beginning of the 18th century, human being has known continued and unrestrained technological development. This industrialisation process caused an uninterrupted consumption of fossil fuels (coal, crude oil and natural gas). Together with the worldwide population growth and the consequent increase of the energy demand, the reserves of fossil fuels reached, nowadays, the so called Hubbert´s peak.[1],[2] In other words these reserves will exhaust sooner or later. Moreover, the combustion of fossil fuels generates carbon monoxide, nitrogen oxides and not reacted hydrocarbons which cause sincere environmental problems. [3],[4] Greenhouses gas emissions increased to values that overtake the concentrations regularly processed by nature (by plants and ocean).[5] Studies clearly demonstrate that the overproduced amount of carbon dioxide in the atmosphere contributed to rise up the average temperature on earth. [3],[4] A very recent research links the increasing greenhouse-gases concentrations with the growing intensity of rain and snow in the Northern Hemisphere and the higher risk of flooding in the United Kingdom.[6],[7]

In 1970, when this irreversible situation was first realised, a reliable concept of clean Hydrogen Economy started to be implemented. An economy based on hydrogen represents a remarkable step towards the independence of a carbon energy vector. In addition, since 1984, there is a growing gap between new oil source discoveries and oil production which forbids, nowadays, to think about an oil based economy.[8] Hydrogen is regarded as a suitable energy carrier due to its high abundance and low weight.[5],[9] It forms a clean oxidation product (water) and it shows the highest energy value per mass of all chemical fuels. Unlike fossil fuels it is considered a secondary energy source because it needs to be synthesised using energy. However, hydrogen could be produced by renewable energy sources (solar and/or wind energy) contributing to decrease the environmental pollution and to increase the security of energy supply.[5] Figure 1.1 represents an ideal hydrogen cycle where it is produced by water splitting (by means of solar energy), confined safely and reversibly in a solid material and used to feed a fuel cell to produce electrical energy.[5]

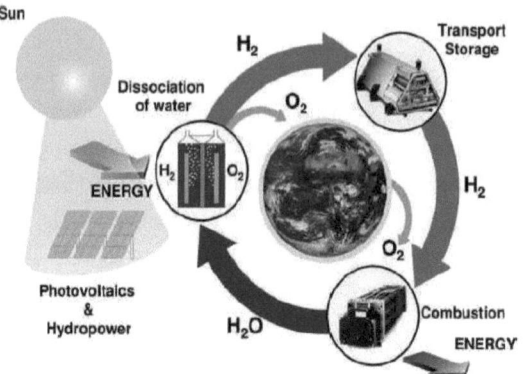

Figure 1.1. Hypothetical hydrogen cycle. (Reproduced from reference 5)

Concerning on-board hydrogen storage, nowadays, there are mainly three technologies: compressed gas cylinders, liquid hydrogen tanks and solid state storage.[5]

In compressed gas cylinders, hydrogen is pressurised to values that lie in the range of 200-800 bar. Stainless steel and composite materials are used for vessels able to withstand low pressure and high pressure values respectively. Unfortunately, an increase of the hydrogen pressure causes a decrease in the gravimetric hydrogen capacity due to the thicker tank walls.[5] This technology suffers from a comparably low volumetric density and high cost of gas compression. In addition, safety issues are matter of concern. In order to overcome these drawbacks, future tanks are planned to be constituted by three layers: an inner polymer liner, a stress withstanding carbon-fibre component and an outer layer of aramid-material (a fibre) against mechanical and corrosion damages.[5] This design should be able to meet the requirements settled by industries which are 700 bar tanks with 110 kg mass.

Liquid hydrogen is confined in tanks at 21.2 K and 1 bar pressure. Due to the critical temperature of hydrogen (33 K), it can only be stored in open systems. In a closed one, it would reach pressure values up to 104 bar at room temperature.[5] The challenges that this technology offers are represented by the low energy efficiency of the liquefaction process as well as the thermal insulation of the tank in order to limit the boil off phenomenon. In fact, a loss of hydrogen of 2-3 % per day is estimated.[5]

A younger technology, named "cryo-compressed" seems to be very promising.[10] The term "cryo-compressed" indicates storage of hydrogen performed at cryogenic temperatures and high pressures. In this case, hydrogen can be stored as liquid or cold compressed gas. One of the main advantages represented by this technology is that the boil-off phenomenon is drastically reduced because the high storing pressure can be reached before the vent valve activates. In this case, the volumetric system capacity (32 g/L) is higher compared to the other storage solutions proposed so far.[10] Since hydrogen is refuelled at ambient temperature, the fuelling stations costs are reduced. In addition, costs for cryo-compressed technology are reduced of 50 % and 20 % compared to 700 and 350 bar compressed gas systems respectively.[10] Moreover, the volumetric efficiency is twice and 40 % higher than 350 and 700 bar system respectively.[10] So far, predictions say that cryo-compressed technology could offer a better capacity respect to the current chemical hydride systems bypassing, in addition, material regeneration concerns.[10]

Compressed gas technology is a complicated technique because of the risk of the gas compression (Japan has prohibited these vessels on the roads for normal vehicles). In addition, the vessels need to be equipped with extra pressure controllers since the hydrogen, to become available, drops from 450 bar to zero over-pressure. Furthermore, when full, these containers would contain only 4 % H_2 by mass.[11] Liquid hydrogen would, in principle, represent an interesting solution because the mass of hydrogen per volume of container can be increased respect to the compressed gas technology. However, the condensation temperature of hydrogen at 1 bar is –252 °C and its critical temperature is –241 °C (above this temperature hydrogen is gaseous).[11] Therefore, hydrogen needs to be stored in open systems in order to prevent overpressure. As consequence, heat transfer through the container leads directly to the loss of hydrogen. Since compressed gas and liquid hydrogen storage technologies are inconvenient solutions for automotive applications due to the abovementioned reasons, another functional solution has to be proposed. Solid state hydrogen storage provides the highest volumetric density among the aforementioned technologies.[5, 9] In addition, due to the low pressure values involved and the endothermic desorption process, it represents the safest technology. Generally, the desired hydrogen storage material should possess high storage capacity, mild operating temperature, fast kinetics, low cost, excellent reversibility and low or, even no, toxicity.[5] So far, there is no a single material fulfilling simultaneously all of these requirements. Important hurdles remain: the cost of a safe and efficient production of hydrogen, its storage and the development of the fuel cell technology.[12]

It is well known that hydrogen reacts in combination with several metallic elements to form metal hydrides (MH_x).[5] A brief chronological evolution of hydrogen storage materials which represents some of the key steps of the achieved progress is summarised in the following pages.

The first example of a metal hydride is represented by PdH_x.[13] In this compound, the hydrogen is not covalently bonded to the metal but occupies interstitial sites within the palladium host lattice. The high solubility and mobility of hydrogen in Pd at room temperature made the Pd–H system one of the most studied and characterised. Further development of this material was hampered by the prohibitive cost of palladium.

With the aim to understand the embrittlement effect that hydrogen causes on zirconium-alloys, the $ZrNiH_3$ compound was discovered.[14] It represents the first example of intermetallic hydride. Still, during the same studies on zirconium intermetallics, scientists observed that hydrogen is able to induce a crystal/glass transformation in metallic alloys. The reaction of $Zr_{0.75}Rh_{0.25}$ with molecular hydrogen induces the formation of amorphous $Zr_{0.75}Rh_{0.25}H_{1.14}$ which represents the first amorphous hydride.[15]

At the end of the 1960s, during investigations on AB_5-type intermetallics, the remarkable hydrogen storage properties of these alloys were discovered. For instance, $SmCo_5$ reacts with hydrogen as reported by the following reaction:[16]

$$SmCo_5 + 1.25H_2 \leftrightarrow SmCo_5H_{2.5}$$

Also the $LaNi_5$ alloy offered, at that time, interesting hydrogen storage properties and the hydrogenation reaction is presented herein:[17]

$$LaNi_5 + 3.35H_2 \leftrightarrow LaNi_5H_{6.7}$$

This reaction is reversible at room temperature and, as in case of the Pd, the hydrogen is located interstitially among the metal atoms. Both materials were and are employed as anodes in rechargeable nickel-metal hydrides batteries which combine, within the cell, a negative hydride electrode together with a positive nickel electrode.[18]

After the oil crisis in the 1970s, the interest in solid state hydrogen storage was risen up when Reilly discovered an $MmNi_5$ alloy (Mm = mischmetal).[19] The mischmetal is normally a mixture of rare-earth metals, mostly lanthanum and neodymium. Unfortunately, the main drawback of these materials is represented by their cost. Sandrock[20] showed an improvement of the properties of these AB_5 alloys through partial substitution of the A and B elements with Ca and Al or Ni respectively. This approach was effective in reducing both the equilibrium pressure to 1 or 2 bar at room temperature as well as the cost of production by 30 %.

During studies on AB_5 alloys, Reilly and Wiswall[21] discovered the hydrogen storage properties of the AB-type hydrides. Among them, Fe-Ti hydride was the most studied because it offered better storage capacities (1.5 wt. % H_2 at room temperature) and lower cost compared to the AB_5-type. However, due to the high rate of the heat transfer during absorption/desorption, the sensitivity to O_2, H_2O and CO and their high weight, a future utilisation was limited.

In the early 1980s, the discover of multiphase AB_2-type alloys or, so called, Laves phase with an high degree of structural disorder was reported by Ovshinsky.[22] The hydrogen storage properties of these multicomponent alloys were improved when compositional and structural disorder was introduced. The most representative were ZrV_2, $ZrMn_2$ and $TiMn_2$.

The hydrogen storage capacities offered by the intermetallic hydrides were too low to be considered for automotive applications. Therefore, the research moved towards materials with higher hydrogen gravimetric density. MgH_2 represented one of the most interesting. In the last decade it was extensively studied due to the high hydrogen capacity of 7.6 wt. %, the low cost and availability. Unfortunately, high thermodynamic stability, sluggish sorption kinetics and high sensitivity to oxygen still constitute relevant drawbacks. The thermodynamic stability of MgH_2 results in hydrogen desorption temperature of 300 °C at 1 bar H_2 pressure. This value is still well far from the requirements set for automotive applications.

A critical factor that negatively influences the hydrogenation kinetics of Mg is the presence of a MgO layer on the surface of the metal. This layer creates a barrier against the diffusion of hydrogen and therefore to the formation of magnesium hydride. Recently, this issue was solved by Barkhordarian *et al.*[23] who showed that the addition of Nb_2O_5 additive to MgH_2 remarkably improves its sorption kinetics. MgH_2 milled with 0.5 mol % of Nb_2O_5 desorbs and absorbs hydrogen at 300 C within less than two minutes. The multivalency of the Nb metal cation and the electronic exchange reactions with hydrogen molecules were proposed to be the reasons of the kinetic enhancement. [23] Schimmel *et al.*[24] suggest that the cubic Nb-Mg-O perovskite phase resulting from alloying of Nb and the MgO (present on the surface of the Mg particles although the starting materials are handled in inert atmosphere) might be catalysing the sorption reactions by splitting and transport of hydrogen. Currently, complex hydrides are considered ideal candidates for solid state hydrogen storage due to both their high hydrogen content in volume and weight. These materials are characterised by a central atom which is covalently bonded to the hydrogen. Tetrahydroaluminates (or alanates $[AlH_4]^-$)[25], amides ($[NH_2]^-$)[26] and tetrahydroborates (or borohydrides $[BH_4]^-$)[27] are important examples of this category. Even though tetrahydroborates of the alkali and alkali earth metals exist for over 50 years only a few information about their physical properties are known. This lack of interest could be addressed to the drastic conditions required for reversible hydrogenation. In fact, their high thermodynamic stability and, in many cases, high kinetic barriers for hydrogen uptake and release are reflected by high hydrogen desorption temperatures.[28] Figure 1.2 reports the gravimetric and the volumetric hydrogen densities of several hydrogen storage materials. The Figure highlights the superior hydrogen concentration of some complex hydrides compared to the traditional metal hydrides.

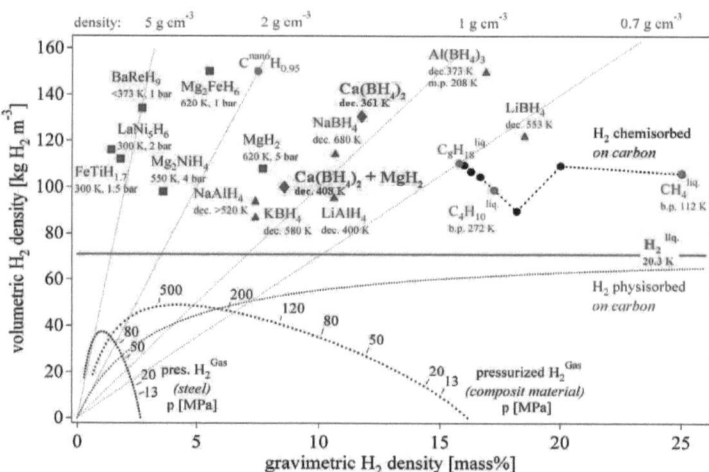

Figure 1.2. Gravimetric density over volumetric density for selected hydrogen storage materials. (Reproduced from reference 5)

Unlike interstitial hydrides, in many cases, the hydrogen desorption of complex hydrides often leads to the formation of at least two compounds. The following scheme of reaction defines the common decomposition path for complex borohydride (M = metal).[5]

$$MBH_4 \rightarrow MH + B + 3/2 H_2$$

Although, in principle, some of them show suitable thermodynamics for on-board applications, their sluggish (re)absorption kinetics is regarded as a severe drawback. The kinetic barrier is represented by the inertness of boron. It is still not clear, whether this behaviour originates from the specific modification of boron (passivation) or from intrinsic properties of the material itself.

Alkaline earth tetrahydroborates are characterised by a thermodynamic stability between ionic and covalent. Group III and transition metal tetrahydroborates are covalent bonded and are either liquids or solids which might sublimate. (Ti(BH$_4$)$_3$ and Zr(BH$_4$)$_4$ decompose at 25 °C).[5] The stability of metal tetrahydroborates is related to their portion of ionic character and those compounds with less ionic character than B$_2$H$_6$ are expected to be unstable.[5] The tetrahydroaluminates are more unstable than tetrahydroborates and therefore more reactive. The difference of stability between tetrahydroaluminate and tetrahydroborates is due to the different Pauling electronegativity of B and Al (2.04 and 1.61 respectively).[5] However, the physical properties of tetrahydroborates and tetrahydroalanates are to a large extent still not known.

The scenario changed when, in 1997, Bogdanovic and Schwickardi showed the positive effect of catalytic amount of Ti-based compounds on the kinetics and the reversibility of NaAlH$_4$.[25] This discovery stimulated intensive investigation of complex hydrides as practical materials for hydrogen storage by the scientific community.

Due to promising thermodynamic properties and a rather high gravimetric storage capacity of up to 5.5 wt. % of hydrogen, NaAlH$_4$ is one of the most investigated compounds among the complex hydrides. The two subsequent decomposition steps are reported below:[29]

(1) 3NaAlH$_4$ ↔ Na$_3$AlH$_6$ + 2Al + 3H$_2$ (ΔH = 3x37 kJ mol^{-1}; 3.7 wt. % H)

(2) Na$_3$AlH$_6$ ↔ 3NaH + Al + 3/2H$_2$ (ΔH = 47 kJ mol^{-1}; 1.8 wt. % H)

Non-milled NaAlH$_4$ melts at 183 °C but the hydrogen desorption starts at 240 °C. The release of hydrogen occurs together with the formation of Na$_3$AlH$_6$ and Al (reaction 1). At temperatures above 300 °C, the respective Na$_3$AlH$_6$ decomposes releasing hydrogen and forming NaH and Al (reaction 2). On the other hand, milled NaAlH$_4$ and the corresponding Na$_3$AlH$_6$ desorb hydrogen below the melting temperature. The reason for this might be the introduction of additional defects like grain boundaries as well as additional interphase areas.[29]

Another promising hydrogen storage system is represented by the Li$_3$N. It can reversibly store hydrogen following a two-step reaction (3):[26]

(3) $Li_3N + 2H_2 \rightarrow Li_2NH + LiH + H_2 \leftrightarrow LiNH_2 + 2LiH$

A recent study showed that a mixture of $LiNH_2$ and LiH can be produced by reactive milling of Li_3N at 20 bar H_2 pressure in only 4 hours.[30]

The enthalpy associated with reaction (3) is -96.3 kJ mol^{-1} H_2[31] and the amount of hydrogen that can be theoretically stored is 10.4 wt. %.[31] The first step of reaction (3) leads to the formation of lithium imide (Li_2NH) and lithium hydride whereas the second leads to the formation of lithium amide ($LiNH_2$) and a further lithium hydride molecule. Given the favourable reaction enthalpy value (-45 kJ $mol^{-1}H_2$)[31] and the high hydrogen capacity (6.5 wt. %), the reversible hydrogenation/(de)hydrogenation reaction between lithium imide and amide (4) is considered to be attractive and competitive with other hydrogen storage systems:

(4) $Li_2NH + H_2 \leftrightarrow LiNH_2 + LiH$

Reaction (4) proceeds under milder thermodynamic conditions (-45 kJ $mol^{-1}H_2$ vs. -96.3 kJ mol^{-1} H_2) compared to reaction (3) (i.e. Li_3N to $LiNH_2$) which requires high vacuum and higher hydrogen desorption temperatures. Chen et al.[32] suggested the decomposition of lithium amide in the presence of LiH to proceed via direct combination of H^+ and H^- from $LiNH_2$ and LiH respectively. Ichikawa et al.[31] identified the same reaction (4) as a two-step process:

(5) $2LiNH_2 \rightarrow Li_2NH + NH_3$

(6) $NH_3 + LiH \rightarrow LiNH_2 + H_2$

The first step of reaction (5) is the decomposition of lithium amide to lithium imide and ammonia in an endothermic reaction. In the second step (6), the ammonia reacts with lithium hydride to produce lithium amide and molecular hydrogen (ΔH = -77 kJ $mol^{-1}H_2$). This latter reaction occurs within 10 min at 230 °C and proceeds continuously until all lithium hydride and lithium amide are consumed.[33]

David et al.[34] rationalised hydrogenation and (de)hydrogenation in the imide-amide system via a mechanism involving the generation of cation vacancies. According to him, during cycling, hydrogenation and (de)hydrogenation occur via the formation of several non-stoichiometric intermediates based on the Li_2NH antifluorite structure. During hydrogenation, the creation of Li$^+$ vacancies in the imide structure is compensated by the formation of new N-H bonds (lithium amide). During (de)hydrogenation, the key mechanism is the lithium migration to form [LiLiNH$_2$]$^+$ and [☐NH$_2$]$^-$(☐=vacancy) species which can bond to hydrogen.[34] Unfortunately, the potential application of these materials is restrained by the high kinetic barrier and the release of poisonous gas (NH$_3$). A recent work published by Hino et al.[35], reports about the amount of ammonia evolved by a sample of LiH + LiNH$_2$. The concentration of NH$_3$ is ca. 0.1 % of that of H$_2$ (above 275 °C).[35] This value would be enough to damage a polymer electrolyte fuel cell (PEFC). Rajalakshmi et al.[36] observed a severe decrease of the performances of the PEFCs when the concentration of NH$_3$ was higher than 10 ppm. Uribe et al.[37] indicated NH$_4^+$ (product of the reaction between H$^+$ and NH$_3$) as responsible of the decreased performances of the PEFC. Therefore, it is essential to find a way to trap or suppress the released ammonia if this system ought to be utilised for hydrogen storage in combination with a PEM fuel cell.

Several strategies were adopted to improve the dehydrogenation performances of too stable hydrogen storage materials. Thermodynamic destabilisation, cation/anion substitution, catalytic activation and nanoconfinement are some examples.

Several strategies were applied to tune reaction enthalpies of metal hydrides and especially of MgH$_2$.[38] The introduction of destabilising agents is used to favour new decomposition paths that can stabilise the (de)hydrogenated products and destabilise the hydride reactants.[38] Figure 1.3 depicts the generalised concept.

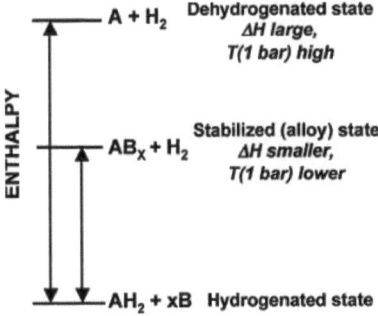

Figure 1.3. Enthalpy diagram depicting the thermodynamic destabilisation through alloy formation upon dehydrogenation. (Reproduced from reference 38).

The first clear example of thermodynamic destabilisation was discovered by Reilly and Wiswall for the Mg-Ni hydrogen storage system.[39] In this system, Mg_2Ni reacts with hydrogen producing a ternary hydride: Mg_2NiH_4. The reaction is reversible and upon decomposition the starting materials are reversibly formed.

In the (de)hydrogenated state, Mg_2Ni is stabilised by -21 kJ mol^{-1} Mg compared to pure Mg.[9] In the hydrogenated state, Mg_2NiH_4 is stabilised by -10 kJ mol^{-1} Mg respect to pure MgH_2.[9] In total, the hydrogen reaction enthalpy of Mg_2Ni is reduced by 11 kJ mol^{-1} H_2 ($\Delta H_{(Mg2Ni-H)}$ = 67 kJ mol^{-1} H_2).[9] Mg_2NiH_4 exhibits a plateau pressure at ca. 240 °C.[9] Klassen et al.[40] showed that the plateau pressure is reduced to 230 °C in case of alloying and substitution of Ni by Cu ($Mg_2Ni_{0.5}Cu_{0.5}$). However, Mg_2NiH_4 offers only 3.6 wt. % H_2 capacity which is less than half the value of MgH_2.

Reilly and Wiswall[41] showed the change of the thermodynamic properties also for the MgH_2/Mg_2Cu system. In that work, Mg_2Cu was reversibly hydrogenated to $3/2MgH_2$ + $1/2MgCu_2$ with an equilibrium pressure of 1 bar at 240 °C. This temperature value is 40 °C lower than the temperature for pure MgH_2 at 1 bar pressure. Therefore, the total value of reaction enthalpy in this system is successfully lowered by the addition of $MgCu_2$. Later on, Vajo et al.[42] reported the reduction of reaction enthalpy of MgH_2 by adding Si as reacting additive. The interaction between Si and MgH_2 during desorption results into the formation of a stable $MgSi_2$ alloy instead of simple Mg. In this way, the overall reaction enthalpy is decreased from 75.3 kJ mol^{-1} H_2 to 36.4 kJ mol^{-1} H_2.[42] Vajo et al.[42] applied the same concept to the LiH. In this case, the addition of Si leads to the formation of Li_4Si which results in a decrease of the total reaction enthalpy of this system to 70 kJ mol^{-1} H_2. In principle, the reduction of the total reaction enthalpy value by adding Si is an interesting approach. However, the addition of such an element into the system generates undesirable loss in capacity.

$LiBH_4$ has an extremely high gravimetric hydrogen capacity (18.4 wt. %) that makes it an ideal candidate for solid state hydrogen storage. However, the desorption reaction leads to the formation of lithium hydride (LiH) and boron. The (re)hydrogenation reaction, performed at 690 °C and 200 bar H_2 for over 12 hours, does not lead to the complete formation of $LiBH_4$.[43] In this instance, high pressures, high temperatures and long reaction time are required in order to reversibly form the $[BH_4]^-$ anion.[43] Apart from the unfavourable thermodynamics for the formation of B_2H_6, the kinetic barrier is represented by the inertness of boron.[44] Bösenberg[45] reports that, in case of the absorption reactions (starting from LiH and MgB_2) a contracting volume model is predicted. It seems that, at the beginning of the reaction, the interface plays the main role as a limiting process whereas, later on, due to the increasing diffusion paths, diffusion limitations become more important. Bösenberg[45] suggests that the reactivity and mobility of the boron and boron compounds is regarded as a key drawback for the sorption kinetics. Vajo et al.[38] reported that the combination of MgH_2 and $LiBH_4$ resulted in a reduced reaction enthalpy (from 67 to 41 kJ mol^{-1}) due to the formation of MgB_2 instead of boron after desorption. Independently, Barkhordarian et al.[46, 47] discovered the outstanding kinetic effect of MgB_2. This unexpected effect was highlighted during the attempt to synthesise tetrahydroborates applying the so called RHC concept (RHC = Reactive Hydrides Composite). He found that the kinetic barriers for the formation of $LiBH_4$, $NaBH_4$ and $Ca(BH_4)_2$ are drastically reduced when MgB_2 is used instead of B as starting material. This unexpected kinetic effect enables the absorption reactions of tetrahydroborate to proceed with lower enthalpy of reaction if compared to those starting from the elements. In this way, the energy efficiency of the hydrogen storage system is increased.[46]

Barkhordarian et al.[46] succeeded in forming $LiBH_4$, $NaBH_4$ and $Ca(BH_4)_2$ by gas phase loading, using MgB_2 instead of pure boron as well as the respective alkaline/alkaline earth hydrides as starting materials. The kinetic enhancement is linked to the peculiar layer structure of MgB_2.[46] Parallel formation of MgH_2 as side product was observed in all the reactions which are reported in Table 1.1:

$2LiBH_4 + MgH_2 \rightarrow 2LiH + MgB_2 + 4H_2$ ($\Delta H = 23.0$ kJ mol^{-1} H_2 ; 11.4 wt. % H)[46]

$2NaBH_4 + MgH_2 \rightarrow 2NaH + MgB_2 + 4H_2$ ($\Delta H = 31.0$ kJ mol^{-1} H_2; 7.8 wt. % H)[46]

$Ca(BH_4)_2 + MgH_2 \rightarrow CaH_2 + MgB_2 + 4H_2$ ($\Delta H = 46.9$ kJ mol^{-1} H_2 ; 8.3 wt. % H)[48]

Table 1.1. Decomposition paths for some Reactive Hydride Composite Systems.

Another approach to tailor thermodynamics and kinetics was represented by the cation[49] and anion substitution[50-52] within the host structure of a certain hydrogen storage material. With this procedure, the binding energy is modified and hence the (de)hydrogenation enthalpy and kinetics are altered. The cation substitution represents a more feasible process due to the large variety of metal cations. Miwa et al.[49] reported that the partial substitution of Li^+ with more electronegative cations (Cu^+) within $LiBH_4$ decreases its thermodynamic stability. A similar approach was performed with success on $LiNH_2$.[53]

On the other hand, anion substitution is also effective in tailoring thermodynamics. Kang et al.[52] presented an increase of the plateau pressure of Na_3AlH_6 doped with TiF_3. The thermodynamic destabilisation is attributed to the F⁻ substitution within the hydrogen sublattice.[50-52] Yin *et al.*[54], by DFT calculations, confirm the F⁻ substitution within LiBH$_4$. Yin *et al.*[55], through the combination of computational work and observations gained by experiments, established a "functional anion" concept. In addition, their calculation show that F⁻/H⁻ substitution for tetrahydroborates is expected to be more pronounced if compared to alanates.[54]

Concerning kinetics, a positive enhancement can be attained modifying grain and particle sizes to the nanoscale range.[56, 57] This approach is effective in altering the kinetic barrier of the mass transport reducing the diffusion distance, increasing the diffusion rate and facilitating nucleation.[58, 59] However, particle agglomeration and growth and in some cases sintering, are inevitable due to the high surface energy involved upon cycling. These phenomena lead to an irreversible degradation of the performance of the material. Nanoconfined materials (nanoscaffolds)[57, 59] simultaneously provide retention of both nanostructure and morphology upon cycling. Concerning applications of the concept on complex hydrides, NaAlH$_4$ and LiBH$_4$ were extensively used.[56, 57] In case of LiBH$_4$ confined in aerogel, Gross *et al.*[59] observed enhanced sorption kinetics but, due to experimental uncertainties, they were unable to determine any thermodynamic destabilisation although the interaction of LiBH$_4$ with the pore walls, during desorption, might have changed its surface energy, increasing therefore the equilibrium pressure. This approach mainly improves the interfacial contact and the interactions among the reacting phases.[59]

Unfortunately, the introduction of inert carbon into the system leads to a considerable capacity loss. In addition, this strategy is still limited by melting point, solubility and wetting property of the hydrogen storage material.[56]

1.1 Ca(BH$_4$)$_2$

Among the aforementioned light metal tetrahydroborates, Ca(BH$_4$)$_2$ represents a potential candidate for solid state hydrogen storage due to its high gravimetric (11.5 wt. %) and volumetric (~130 kg m^{-3}) hydrogen content.[60] Furthermore, the (de)hydrogenation enthalpy was recently calculated to be 32 kJ mol^{-1}H$_2$[61, 62] if CaH$_2$ and CaB$_6$ are the decomposition products which is within the optimal range for mobile applications.[61, 62] This would reflect in a decomposition temperature lower than 100 °C at 1 bar H$_2$ pressure.

The following decomposition paths for pure Ca(BH$_4$)$_2$ are discussed in literature:

(7) Ca(BH$_4$)$_2$ ↔ 2/3 CaH$_2$ + 1/3 CaB$_6$ + 10/3 H$_2$
(8) Ca(BH$_4$)$_2$ ↔ CaH$_2$ + 2 B + 3 H$_2$
(9) Ca(BH$_4$)$_2$ ↔ CaB$_2$H$_2$ + 3 H$_2$
(10) Ca(BH$_4$)$_2$ ↔ CaB$_2$H$_6$ + H$_2$
(11) Ca(BH$_4$)$_2$ ↔ 1/6 CaB$_{12}$H$_{12}$ + 5/6 CaH$_2$ + 13/6 H$_2$

Besides CaH$_2$ and H$_2$, different boron compounds are reported in literature: CaB$_6$, B (boron), CaB$_2$H$_2$, CaB$_2$H$_6$ and CaB$_{12}$H$_{12}$.[63-66] The calculated enthalpies of reaction are 37.04[66], 57.3[48], 31.09[66] and 39.2[63] or 31.34[66] kJ mol^{-1} H$_2$ for reaction (7), (8), (10) and (11) respectively. These values are calculated at 300 K and 1 bar H$_2$ pressure. For reaction 9, Zhang et al.[66] report a reaction enthalpy value of 68.51 kJ mol^{-1} H$_2$, calculated at 0 K ignoring the zero-point energy.

Species containing [B$_{12}$H$_{12}$]$^{2-}$ were predicted to be likely during decomposition of tetrahydroborates[63] and their chemical stability is known to be rather high.[67] However, their detection is difficult. The existence of several amorphous polymorphs of CaB$_{12}$H$_{12}$ during (de)hydrogenation reaction of Ca(BH$_4$)$_2$ (reaction 11) was predicted by Wang et al.[65] These phases have competing enthalpies of reaction (ΔH^{0K} = 35.8–37.9 kJ mol^{-1}H$_2$).[65]

Recently, Riktor et al.[68] and Lee et al.[69] observed the formation of CaB$_2$H$_x$ (x = 2) (reaction 9) and of CaB$_m$H$_n$ phase respectively. Zhang et al.[66] found the phase proposed by Riktor et al.[68] too unstable to be a decomposition product (ca. 50 kJ mol^{-1} H$_2$ > CaB$_2$H$_6$ and CaB$_{12}$H$_{12}$). DFT and PEGS (Prototype Electrostatic Ground-State) calculations performed by Zhang et al.[66] showed CaB$_2$H$_6$ to be more likely (reaction 10). When vibrational entropy and free energy (including ZPE, zero-point energy) are taken into account, the reaction enthalpy value of CaB_2H_6 (reaction 10) competes in energy with that of $CaB_{12}H_{12}$ (reaction 11) within 1 kJ mol^{-1} H$_2$.[66] However, this value is smaller than the limit of accuracy of the DFT method itself.

Experimentally, the thermal decomposition reaction of calcium borohydride involves two (de)hydrogenation steps. The first decomposition step starts around 350 °C and leads to the formation of CaH$_2$ and an unknown intermediate phase which decomposes further, in the second step, in the temperature range of 390-500 °C.[70]

In the aforementioned works, X-ray diffraction was mainly employed to detect the existence of intermediate phases but, because some of them might be in the amorphous state, diffraction does not always represent the proper tool. Since $^{11}B\{^1H\}$ Solid State Magic Angle Spinning- Nuclear Magnetic Resonance (MAS-NMR) does not suffer this limitation, it will be used in this study.

Ca(BH$_4$)$_2$ exists as several structural polymorphs. First of all, there are two low temperature modifications, α (space group F2dd) and γ (Pbca), recently indexed as orthorhombic phases.[60, 71] These room temperature structures transform into a high temperature phase β, in the 180-300 °C temperature range.[71] Another polymorph, called α' with tetragonal cell (I-42d) was reported to form at 222 °C.[71] A polymorph, called δ, was found upon heating together with a not yet indexed phase, still stable at 500 °C.[64] Recently, Riktor *et al.* identified the δ phase to be a calcium borohydride borate with composition Ca$_3$(^{11}BD$_4$)$_3$(^{11}BO$_3$).[72] This phase is the result of a partial oxidation reaction. Its stability was confirmed by DFT calculations.[72]

The formation of Ca(BH$_4$)$_2$ was shown to be partially reversible by using suitable additives. Ronnebro *et al.*[73] were able to synthesise Ca(BH$_4$)$_2$ with a yield of 60 % from a mixture of CaH$_2$ and CaB$_6$ with Pd and TiCl$_3$ by applying 700 bar H$_2$ and temperatures of 400-440 °C. Rongeat *et al.*[74] showed that 19 % of calcium borohydride was obtained by high-pressure reactive ball milling (near room temperature) of CaH$_2$ and CaB$_6$ after 24 hours at 140 bar H$_2$ employing TiF$_3$ or TiCl$_3$ as additives. This yield was improved to 60 % during further cycling of the material at 350 °C and 90 bar H$_2$ for 40 hours. Lately, some TM-fluorides and chlorides (TiCl$_3$ and NbF$_5$) have demonstrated to positively affect its partial reversible formation. Over 50 % hydrogen can be reversibly absorbed when 90 bar H$_2$ pressure and 350 °C are applied for 24 hours to the decomposition products (Ca-H-Cl and boron or CaB$_6$ in case of TiCl$_3$; CaF$_{2-x}$H$_x$ and CaB$_6$ in case of NbF$_5$) of calcium borohydride catalysed by TiCl$_3$ or NbF$_5$.[75, 76] Only partial formation of Ca(BH$_4$)$_2$ could be obtained so far and no detailed explanation concerning the reaction mechanism was provided due to the complexity of the system itself and to the sensitivity of the powder to both the moisture and the microscope electron beam. All these limitations invoke the simultaneous application of several experimental methods and of their corresponding results.

1.2 $Ca(BH_4)_2 + MgH_2$

As already reported in section 1, the Reactive Hydride Composites (RHC) concept represents an advantageous approach due to the possibility of tuning the reaction thermodynamics by choosing appropriate reactants. However kinetic restrictions have to be considered.

With both a theoretical hydrogen storage capacity of 10.5 wt. % and an estimated equilibrium temperature <160 °C [46], $Ca(BH_4)_2 + MgH_2$ composite represents a promising candidate for mobile hydrogen storage. The decomposition paths proposed in literature for this system are the following:

(12) $Ca(BH_4)_2 + MgH_2 \leftrightarrow CaH_2 + MgB_2 + 4 H_2$
(13) $Ca(BH_4)_2 + MgH_2 \leftrightarrow 2/3\ CaH_2 + 1/3\ CaB_6 + Mg + 13/3\ H_2$
(14) $Ca(BH_4)_2 + MgH_2 \leftrightarrow CaH_2 + 2 B + Mg + 3 H_2$

The reactions involving formation of MgB_2 (12) or CaB_6 (13) upon hydrogen desorption should be the most thermodynamically favourable because the borides are exothermically formed. Kim et al.[48] reported the existence of a subtle competition between reaction 12 and 13 in dependence of the experimental conditions applied. Taking into account corrected values for the enthalpy of formation of both CaB_6 and MgB_2, SGTE calculations predict that, at 350 °C and 1 bar H_2 [48], formation of MgB_2 is likely whereas 350 °C and dynamic vacuum lead to the formation of CaB_6. However, upon desorption of $Ca(BH_4)_2 + MgH_2$, MgB_2 is not necessarily formed.[77] Kim et al.[48] reported CaH_2, Mg and CaB_6 to be the decomposition products. The (re)absorption reaction, at 90 bar H_2 and 350 °C for 24 hours, led to the formation of 60 % of $Ca(BH_4)_2 + MgH_2$ thus evidencing the important role of CaB_6 for reversibility.[48] The formation of hexaboride (CeB_6 and CaB_6)[78] was already reported to promote reversible hydrogenation reactions ($6LiBH_4 + MH_2$, M = Ca, Ce) although metal-hexaborides are known to be highly stable. Barkhordarian et al.[77] estimated a standard enthalpy value of 27.5 kJ mol^{-1} H_2 for the reaction involving MgB_2 and CaH_2 as decomposition products (reaction 12). No details concerning temperature and pressure values are reported. By DFT (Density Functional Theory) method, Kim et al.[48] calculated it to be

46.9 kJ mol^{-1} H$_2$. DFT calculations for the decomposition reaction leading to CaH$_2$, CaB$_6$ and Mg (reaction 13), indicate a reaction enthalpy value of 45 kJ mol^{-1} H$_2$.[48] If boron is formed instead of CaB$_6$ (reaction 14), the calculated reaction enthalpy value is 57.9 kJ mol^{-1} H$_2$.[48] The enthalpy values of 46.9, 45 and 57.9 kJ mol^{-1} H$_2$ are calculated at 25 °C and 1 bar H$_2$ pressure. The decomposition paths involving formation of MgB$_2$ and CaB$_6$ upon hydrogen desorption should be the most thermodynamically favourable because the borides are exothermically formed.

A combination of X-ray diffraction and ^{11}B{^{1}H} Solid State Magic Angle Spinning-Nuclear Magnetic Resonance will be employed in this study for a detailed characterisation of both sorption mechanism and final decomposition products.

1.3 Aim of the work

In the present work, the sorption properties of the Ca(BH$_4$)$_2$ and Ca(BH$_4$)$_2$ + MgH$_2$ composite system are investigated in detail. The pure system offers a reversible hydrogen storage capacity of 11.5 wt. % and a theoretical (de)hydrogenation enthalpy of 32 kJ mol^{-1}H$_2$ if CaH$_2$ and CaB$_6$ are the decomposition products, which corresponds to an estimated equilibrium temperature of ca. 100 °C at 1 bar H$_2$. The composite system has a theoretical hydrogen storage capacity of 10.5 wt. % and an estimated equilibrium temperature lower than 160 °C. Therefore, the aforementioned systems are ideal for automotive application purposes.

Before this work started only a few data were available to the scientific community about both the chemistry (stability, structure polymorphs or intermediate compounds) and the reaction mechanism of Ca(BH$_4$)$_2$ formation and decomposition. It was known that Ca(BH$_4$)$_2$ could be reversibly formed at 700 bar of H$_2$ pressure (Ronnebro *et al.*) only if a mixture of TiCl$_3$ and Pd was added to the pure starting material. Only recently, the possibility for Ca(BH$_4$)$_2$ to follow multiple decomposition pathways was proposed by computational methods. No experimental evidence was reported. In addition, no information concerning the sorption properties was available.

Concerning the Ca(BH$_4$)$_2$ + MgH$_2$ composite system, the key role played by MgB$_2$ on its formation was reported in the literature (RHC composites). However, no detailed investigation of both the sorption properties and the reaction mechanisms was performed. It was believed that MgB$_2$ was necessary for the reversible reaction to proceed. However, upon desorption, it was not reversibly formed.

A study of the effect played by the addition of transition-metal fluorides additives on the reaction kinetics of the $Ca(BH_4)_2$ and $Ca(BH_4)_2 + MgH_2$ composite system was performed. Transition- and light-metal based additives have shown to be beneficial on the sorption reaction kinetics of the complex hydrides.[23, 25] Transition-metal fluorides are highly reactive. They likely evolve to more stable compounds when combined with a hydride-phase during mechanical treatment or during sorption reactions. Fang *et al.*[79] showed the ball milled $LiBH_4$-TiF_3 mixture to release hydrogen at low temperatures (70–90 °C) without impurities. This improved desorption performance seems to be linked to the simultaneous *in-situ* formation and decomposition of $Ti(BH_4)_3$.[79] Given that $Ca(BH_4)_2$ is less thermodynamically stable than $LiBH_4$, its sorption kinetics could be improved to a greater extent. The influence of the additives on the sorption kinetics is studied by volumetric measurements. The chemical state, size and distribution of the Nb- and Ti-based additives is studied by X-ray absorption spectroscopy. The local structure and the additive/catalyst distribution are presented by Transmission Electron Microscopy (TEM).

An assessment of the role of the additives and of Mg as heterogeneous nucleation sites for the formation of CaB_6 is presented by means of interplanar mismatch. This work should provide an understanding of the mechanism played by the additives and by Mg on the sorption reactions of the $Ca(BH_4)_2$ and $Ca(BH_4)_2 + MgH_2$ composite system.

2 Experimental Part

In the following section an overview of the experimental techniques employed in this study will be provided together with the experimental parameters. Advantages of the methods will be discussed.

2.1 Materials

For investigations reported in sections 3.1 and 3.2, pure $Ca(BH_4)_2$ powder was obtained by drying the commercially available $Ca(BH_4)_2$-2THF adduct (purchased from Sigma-Aldrich) for 1 h and 30 minutes at 200 °C in vacuum and subsequent cooling to room temperature. The process leads to a mixture of low temperature (α and/or γ) and high temperature β-$Ca(BH_4)_2$ polymorphs with different abundance. Figure 2.1 reports the infrared spectra of the samples before and after removal of the solvent thus evidencing the successful reaction.

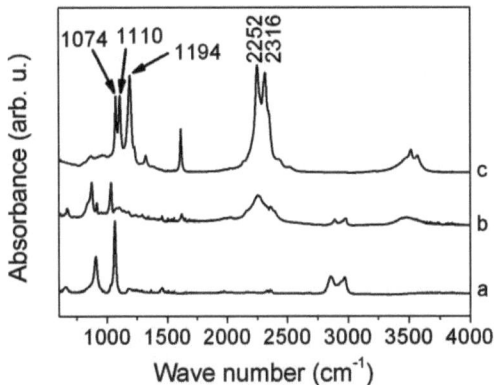

Figure 2.1. FTIR of THF (tetrahydrofuran) (a); $Ca(BH_4)_2$-2THF (b); pure dried $Ca(BH_4)_2$ (c). FTIR measurements were performed at the Institute of Polymer Research at the Helmholtz-Zentrum Geesthacht, Zentrum für Material- und Küstenforschung.

For investigations reported in sections 3.3-3.6, pure $Ca(BH_4)_2$ was purchased from Sigma-Aldrich. The powder contains a mixture of γ and β polymorphs.
The MgH_2 powder (purity 95 %) was purchased from Tego Magnan.
The transition-metal fluoride additives (TM-fluorides) TiF_3 (unknown purity), TiF_4 (purity 98%), VF_3 (purity 98%), VF_4 (purity 95%), NbF_5 (purity 99%) were purchased from Alfa

Aesar as well as the titanium isopropoxide (99.995 % purity). CaF_2 (99.99% purity) was purchased from Sigma-Aldrich.

2.1.1 Chemical Synthesis of $CaB_{12}H_{12}$

The calcium salt of dodecahydrododecaborate dianion was prepared at the Institute de Ciencia de Materiales in Barcelona (Universitat Autònoma de Barcelona, in Bellaterra, Spain) in the group of F. Teixidor Bombardó.

The $Ca[B_{12}H_{12}]$ material was synthesised from the corresponding sodium salt. It was produced with minor modifications of the procedure already reported in literature.[80] In order to exchange the sodium by calcium cation, an aqueous solution of the sodium salt was passed three times over a cation exchange column (cation exchange resin strongly acidic, minimum 2.0 meq/mL), charged with a 3M solution of $CaCl_2$. The produced solution was evaporated and dried in vacuum overnight.

2.2 Sample preparation
2.2.1 $Ca(BH_4)_2$ with additives

Six different samples were prepared adding 0.05 mol of TiF_3 (unknown purity), TiF_4 (purity 98%), VF_3 (purity 98%), VF_4 (purity 95%), NbF_5 (purity 99%), Titanium isopropoxide (99.995% purity) purchased from Alfa Aesar, to $Ca(BH_4)_2$. Another sample was prepared adding 5 wt. % CaF_2 (99.99% purity), purchased from Sigma- Aldrich, to $Ca(BH_4)_2$.

The samples were milled in a stainless steel vial in argon atmosphere for 1 hour and 40 minutes using a Spex Mixer Mill (model 8000) and 14:1 as ball to powder ratio (four spheres of 3.5 gram each one and 1 gram of powder). All powder handling and milling was performed in an MBraun argon box with H_2O and O_2 levels below 10 ppm to prevent contamination.

2.2.2 $Ca(BH_4)_2$ + MgH_2 with and without additives

MgH_2 (Tego-Magnan) was premilled in a stainless steel vial in argon atmosphere for 5 hours using a Spex Mixer Mill (model 8000) and 10:1 as ball to powder ratio (three spheres of 3.5 gram each one and 1 gram of powder). Premilled MgH_2 was added to commercial $Ca(BH_4)_2$ (Sigma-Aldrich) and then further milled in the same stainless steel vial in argon atmosphere for 5 hours using a Spex Mixer Mill (model 8000) and 10:1 as ball to powder ratio (three spheres of 3.5 gram each one and 1 gram of powder). All powder handling and milling was performed in an MBraun argon box with H_2O and O_2 levels below 10 ppm to prevent contamination.

2.3 Kinetic characterisation

Sorption properties and kinetics were evaluated by thermovolumetric measurements using a Sievert-type apparatus designed by Hydro Quebec/HERA Hydrogen Storage System at the Institute of Materials Research of the Helmholtz-Zentrum Geesthacht, Zentrum für Material- und Küstenforschung.

In the case of the $Ca(BH_4)_2$ sample milled with Ti-isopropoxide (Figure 3.28) the experiments were performed in the Hydrogen Lab at the C.S.G.I. - Department of Chemistry - Physical Chemistry Division of the University of Pavia by C. Milanese.

With the term "thermovolumetric measurement" we refer to a volumetric measurement obtained during heating the material. In literature, thermovolumetry is defined as a technique which follows a gas-absorbing or gas-producing reaction by continuously recording the change in the volume of gas consumed or evolved as the material is heated to elevated temperatures at a constant rate.[81] By means of volumetric techniques we can measure the H_2 pressure variation in a reaction chamber. The word "thermo" is specified because the material is continuously heated to elevated temperatures at a constant rate while the volume of gas, consumed or evolved, is recorded.

Concerning $Ca(BH_4)_2$ samples, with and without additives, the milled powders (70-90 mg) were desorbed by heating from room temperature (25 °C) up to 450 °C in static vacuum (0.02 bar the starting pressure value) and subsequently (re)absorbed at 350 °C and 145 bar H_2 for 20 hours (24 hours for the $Ca(BH_4)_2$ sample milled with Ti-isopropoxide (Figure 3.28)). Pure $Ca(BH_4)_2$ was (re)absorbed at 350 °C and 130 bar H_2 for 24 hours.

In case of the $Ca(BH_4)_2 + MgH_2$ samples, the milled powders (120-140 mg) were desorbed by heating from room temperature (25 °C) up to 400 °C in static vacuum (0.02 bar the starting pressure value) and subsequently (re)absorbed at 350 °C and 145 bar H_2 for 24 hours.

The heating rate was 3 °C min^{-1} for both the experiments.

The amount of hydrogen desorbed or absorbed by the samples, is determined by measuring the difference of pressure between a sample holder containing the material and an empty one used as a reference. Since both sample holders have identical design and building material, the hydrogen content of the samples can be calculated using the ideal gas law.

With the purpose of reducing pressure changes during the analysis, an additional 1 liter volume was employed. This volume was kept constantly at the temperature of 40 °C to avoid influences from the surrounding. In case of the $Ca(BH_4)_2$ sample milled with Ti-isopropoxide, about 500 mg of the powder was loaded in the stainless steel sample holder of a PCTPro-2000 apparatus (Setaram & Hy-Energy) under Ar atmosphere in a glove box (MBraun, O_2 and H_2O content < 0.1 ppm) and subsequently subjected to a temperature desorption run (TPD) by heating from room temperature up to 450 °C at 3 °C min^{-1} in vacuum. At the end of the ramp, an isothermal stage was set, long enough (at least 600 min) to guarantee that all the sample mass reach the equilibrium. Subsequently, (re)hydrogenation was performed at 350 °C by reaching 145 bar of hydrogen pressure through three aliquots of 50 bar each. Finally, a second desorption was then performed, in the same conditions of the first one.

2.4 Thermal analysis

Differential Scanning Calorimetry measurements on $Ca(BH_4)_2$ samples (with and without transition-metal fluoride additives) were carried out using a Netzsch STA 409 C in 150 ml min^{-1} argon flow. The analyses were performed at the Institute for Metallic Materials at the Leibniz Institute for Solid State and Materials Research (Dresden). The samples were investigated in the range of 25-500 °C for the samples with additives and up to 550 °C for the pure non-milled calcium borohydride. The heating rate was 5 °C min^{-1} for both the measurements.

The device works in a so-called "dynamic mode". The pressure (3 bar argon) is always constant and is regulated by the gas flow (150 ml min^{-1}). Concerning $Ca(BH_4)_2 + MgH_2$, with and without additives, and $Ca(BH_4)_2$ with Ti-isopropoxide and CaF_2 samples, the calorimetric analyses were performed in a Netzsch STA 409 C in 50 ml min^{-1} argon flow at the Institute of Materials Research of the Helmholtz-Zentrum Geesthacht, Zentrum für Material- und Küstenforschung. The samples were investigated in the range of 25-50 °C with a heating rate of 5 °C min^{-1}.

2.5 Infrared Spectroscopy

Infrared spectroscopy (IR) is a branch of spectroscopy that concerns with the infrared region of the electromagnetic spectrum. Fourier Transform Infrared Spectroscopy (FTIR) is the preferred and commonly used method of infrared spectroscopy. When the infrared radiation hits a sample, some of it is absorbed by the sample itself while some other is transmitted (it passes through). The absorbed energy promotes internal vibrations into the molecule which have characteristic frequencies. The vibrations result in a sequence of peaks within a spectrum which create a unique molecular fingerprint of the sample. The FTIR technique provides qualitative (sequence of frequencies) and quantitative (size of the peaks) information. FTIR spectrometers are able to simultaneously collect data over a wide spectral range. The term "Fourier transform" means that a mathematical transformation (Fourier transformation) is necessary to convert the raw data into the real spectrum.

FTIR measurements were performed at the Institute of Polymer Research at the Helmholtz-Zentrum Geesthacht, Zentrum für Material- und Küstenforschung.

The data were collected using a Bruker Quinox 55 Spectrometer. FTIR spectroscopy, applied after drying the as-received material, confirmed the full removal of the solvent and the presence of calcium borohydride only (Figure 2.1). Furthermore, FTIR was employed to verify the presence of $[BH_4]^-$ after hydrogen absorption.

2.6 *Ex-situ* X-Ray Diffraction

X-ray diffraction represents in material science a fundamental technique to quickly acquire key information (e.g. crystal structures, relative abundance of phases and crystallite domain sizes) on the investigated samples.

The measurements were performed at the Institute for Metallic Materials at the Leibniz Institute for Solid State and Materials Research (Dresden) in collaboration with O. Gutfleisch, at the C.S.G.I. - Department of Chemistry - Physical Chemistry Division of the University of Pavia by C. Milanese ($Ca(BH_4)_2$ + Ti-isopropoxide after 2nd desorption Fig. 3.29) and at both the synchrotron MAX-lab, Lund (Sweden) at the beamline I711[82] in collaboration with T. R. Jensen and at the synchrotron Hasylab, DESY (Hamburg), at the beamline D3.[83]

At the Leibniz Institute, transmission X-ray diffraction measurements were performed in 0.7 mm capillaries on a Stoe Stadi P (Mo $K\alpha_1$) in Debye-Scherrer geometry. The diffractometer is equipped with a curved Ge (111) monochromator and a 6° linear position sensitive detector with a resolution of about 0.06° 2θ at full width-half maximum (FWHM).

At the University of Pavia, the Ca(BH$_4$)$_2$ + Ti-isopropoxide sample after 2nd desorption was measured in reflection in a Bruker D5005 diffractometer (10° ≤ 2θ ≤ 90°, Cu Kα radiation, step scan mode, step width 0.014°, counting time 3 s, 40 kV, 30 mA). A suitable sample holder (Bruker A100B36) was used in order to avoid powder oxidation. In this device, the powders are dispersed on a 20 mm diameter silicon wafer with high-index surface orientation under Ar atmosphere in the glove-box. The Si slice is fixed on a low background plastic disk, which is sealed to a low background dome-like plastic cap by means of a polymeric o-ring.

At the synchrotron, both the beamlines are equipped with a MAR165 CCD detector. Quartz capillaries with 1.0 mm or 0.7 mm outside diameter were filled with powder, closed with candle wax in order to avoid contamination by oxygen or water and then exposed to the beam for 15 or 30 seconds. Longer expositions (240-360 seconds) were performed on some selected samples in order to improve the resolution. The X-ray wavelengths were 0.939 and 0.500 Å for the beamline I711 and D3 respectively.

For comparison purposes, all XRD data are reported referring to the scattering vector 4πsinθ/λ (Å^{-1}).

2.7 *In-situ* Synchrotron Radiation Powder X-ray Diffraction

In-situ Synchrotron Radiation Powder X-ray diffraction (SR-PXD) was carried out at the synchrotron MAX-lab, Lund (Sweden) at the beamline I711[82] in collaboration with T. R. Jensen and at the synchrotron Hasylab, DESY (Hamburg), at the beamline D3.[83] Both the beamlines are equipped with a MAR165 CCD detector. Since the wavelengths were changing during the different experiments, for comparison purposes, all the SR-PXD data are reported referring to the scattering vector 4πsinθ/λ (Å^{-1}).

An especially designed cell for *in-situ* diffraction studies on solid/gas reactions was employed.[84] The cell is able to withstand pressures up to 300 bar and temperatures up to 700 °C. The samples were mounted in a sapphire single crystal tube in an argon filled glovebox with H$_2$O and O$_2$ levels below 0.1 ppm. The temperature was controlled by a thermocouple placed inside the sapphire tube, just next to the sample. A gas supply system was connected to the cell. This allows changing of the gas atmosphere through a vacuum pump contemporaneously to the X-ray data acquisition. The system was flushed with argon and evacuated three times before the valve to the sample was opened prior to the X-ray experiment. The X-ray exposure time was usually 30 s (15 s x 2) per powder diffraction pattern. If not mentioned otherwise, all the experiments were done at a scanning heating rate of 5 K min^{-1}. The FIT2D software was used to remove diffraction sapphire spots from the 2D pictures acquired.

2.8 X-ray absorption spectroscopy

X-ray absorption spectroscopy (XAS) is a powerful technique to determine the local geometry and/or the electronic structure of a given compound.

Each element on the periodic table has a set of unique energy values (absorption edges) corresponding to the different binding energies of its electrons. When the incident X-ray energy matches the binding energy of an electron within an atom of a given sample, the number of X-rays absorbed by the sample increases abruptly, resulting in an absorption edge.

The XAS spectrum is mainly divided in two parts: the pre-edge region called XANES (X-ray Absorption Near Edge Structure) and the one after the edge called EXAFS (Extended X-ray Absorption Fine Structure). The XANES spectrum can provide information on the average oxidation state of the element in the sample. They are also sensitive to the coordination environment of the absorbing atom. The EXAFS spectrum provides information on the geometrical distribution of the surrounding atoms. To match the spectrum of an unknown sample it is fundamental to collect the spectra of known standard compounds. In this work, the EXAFS spectra of the as milled and (de)hydrogenated/(re)hydrogenated materials will be reported together with those of the standard compounds.

XAS measurements of the $Ca(BH_4)_2$ and $Ca(BH_4)_2$ + MgH_2 samples with 5 mol % TiF_4 and NbF_5 were performed in transmission mode at the synchrotron Hasylab, DESY (Hamburg), at the beamline A1 and C respectively. Spectra were collected at the Ti and Nb K-edge (4966 and 18986 eV respectively) under vacuum at ambient temperature.

The beamlines are equipped with a Si (111) channel cut double crystal monochromator. In addition, beamline A1 has a two mirror system (the first mirror is coated with Ni and the second mirror has two stripes, one coated with Ni and the other one uncoated SiO_2) to suppress higher harmonic rejection.

The as milled and (de)hydrogenated/(re)hydrogenated samples were mixed with cellulose powder for dilution before XAS experiments. The mixed powders were pressed to pellet shape and placed between two Kapton foils on aluminium sample holders. This procedure was performed in an argon filled glovebox.

Ti-foil, TiB_2, Ti_2O_3, TiO_2 (anatase), TiO, TiF_3 and TiF_4 samples were used as a reference for XAS measurements of the Ti K-edge. Nb-foil, NbO, NbF_5, NbB_2, Nb_2O_5 and $MgNb_2O_6$

samples were used as a reference for XAS measurements of the Nb K-edge. EXAFS data processing was carried out by the software ATHENA and ARTHEMIS,[85] two interactive graphical utility based on the IFFEFIT[86] library of numerical and XAS algorithms. XANES data analyses were performed subtracting the pre-edge background and normalizing the edge.

2.9 ^{11}B Magic Angle Spinning-Nuclear Magnetic Resonance

X-ray diffraction does not represent the proper tool to detect fine nano-crystalline and/or amorphous compounds. With the purpose of identifying the nature of the final B-containing compounds in the desorption products of $Ca(BH_4)_2$ and $Ca(BH_4)_2$ + MgH_2 system, $^{11}B\{^1H\}$ MAS-NMR measurements were carried out.

The one dimensional ^{11}B Magic Angle Spinning-Nuclear Magnetic Resonance ((1D) $^{11}B\{^1H\}$ MAS-NMR) measurements were performed at the Servei de Ressonància Magnètica Nuclear at the Universitat Autònoma de Barcelona, Bellaterra (Spain) in collaboration with M. D. Baró and P. Nolis.

Solid state MAS-NMR spectra were obtained using a Bruker Avance 400 MHz spectrometer with a wide bore 9.4 T magnet and employing a boron-free Bruker 4 mm CPMAS probe. The spectral frequency was 128.33 MHz for the ^{11}B nucleus and the NMR shifts are reported in parts per million (ppm) externally referenced to BF_3Et_2O. The powder materials were packed into 4 mm ZrO_2 rotors in an argon-filled glove box and were sealed with tight fitting Kel-F caps. The one dimensional (1D) $^{11}B\{^1H\}$ MAS-NMR spectra were acquired after a 2.7 μs single π/2 pulse (corresponding to a radiofield strength of 92.6 kHz) and with application of a strong ^{1}H signal decoupling by using the two-pulse phase modulation (TPPM) scheme. The spectra were recorded at a MAS spinning rate of 12 kHz. Sample spinning was performed using dry nitrogen gas. The recovery delay was set to 10 seconds. Spectra were acquired at 20 °C (controlled by a BRUKER BCU unit).

2.10 Transmission Electron Microscopy

The materials were analysed at the Servei de Microscòpia at the Universitat Autònoma de Barcelona, Bellaterra (Spain) by E. Pellicer and E. Rossinyol. TEM characterisation, including imaging and selected area electron diffraction (SAED), was performed on a Jeol-JEM 2011 microscope operated at 200 kV. The powders were dispersed in THF under Ar atmosphere in a glove box and then a couple of drops were placed dropwise onto a holey carbon supported grid. To prevent from oxidation, the sample holder was immediately inserted into the microscope. Elemental chemical analyses were performed with an Energy-Dispersive X-ray Spectroscopy (EDX) system coupled to the TEM microscope. As a consequence of the experimental setup, Cu and Cr peaks, belonging to the TEM grid, and the Si signal, belonging to the sample holder, might be included in the EDX spectra.

For the dark field image, the following procedure was used: when a micrograph is formed in the TEM microscope, the central transmitted spot is normally used to obtain the conventional bright field images. However, if an aperture is inserted into the back focal plane of the objective lens to block the central spot, the micrograph is obtained from the scattered electrons. These images are called dark-field. In the direction of the selected aperture, they appear dark where no electrons scattering occur, slightly grey in amorphous areas and bright in zones where many electrons scatter.

3 Results

The sorption properties of the $Ca(BH_4)_2$ and $Ca(BH_4)_2$ + MgH_2 Reactive Hydride Composite system, with and without additives, were studied by means of volumetric measurements, calorimetric techniques, *in* and *ex-situ* XRD, X-ray absorption spectroscopy, ^{11}B Magic Angle Spinning-Nuclear Magnetic Resonance and Transmission Electron Microscopy. This chapter will report about the hydrogen sorption reaction of the pure non-milled $Ca(BH_4)_2$, followed by the investigation of the effect of several additives (transition metal fluorides, Ti-isopropoxide and CaF_2) on its reversible formation. Section 3.5 and 3.6 will present the results for the $Ca(BH_4)_2$ + MgH_2 composite system and the influence of some of the aforementioned additives on its reversible formation respectively.

3.1 $Ca(BH_4)_2$

This section presents the results on the pure non-milled $Ca(BH_4)_2$ system.

3.1.1 The (De)hydrogenation Reaction

The volumetric analysis for the first hydrogen desorption reaction of the pure non-milled $Ca(BH_4)_2$ system was performed in a Sievert´s type apparatus heating from room temperature (25 °C) up to 450 °C with a constant heating rate of 3 °C min^{-1} under vacuum (0.02 bar the starting pressure value).

The XRD pattern of the starting Ca(BH$_4$)$_2$ material is reported in Figure 3.1. The Figure indicates the presence of the low temperature α (space group F2dd, orthorhombic phase)[71] and of the high temperature β modification (space group P4, tetragonal phase).[60] The narrow reflections in Figure 3.1 define the crystalline status of both the Ca(BH$_4$)$_2$ polymorphs.

The relative phase abundance, calculated by Rietveld method (by MAUD software)[87] are 41 wt. % (± 5 error) and 59 wt. % (± 5 error) for α- and β-phase respectively.

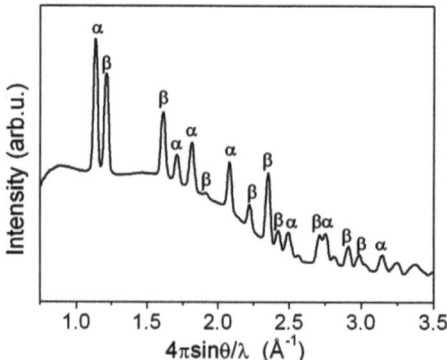

Figure 3.1. SR-PXD pattern at room temperature of Ca(BH$_4$)$_2$. α-Ca(BH$_4$)$_2$ (α); β-Ca(BH$_4$)$_2$ (β). The measurement was performed at the synchrotron MAX-lab, Lund (Sweden) at the beamline I711.

The volumetric analysis for the first hydrogen desorption reaction is reported in Figure 3.2. The kinetic curve evidences two different slopes referred to two distinct desorption steps. The first hydrogen desorption reaction starts at ca. 350 °C whereas the second begins at ca. 390 °C.

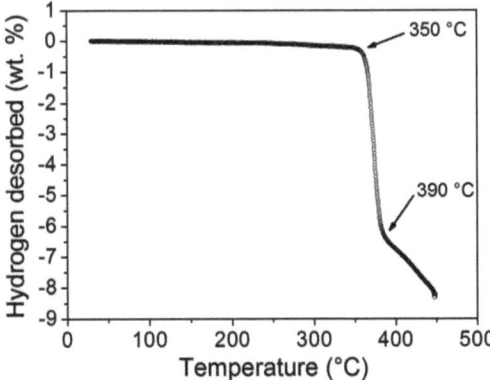

Figure 3.2. Volumetric measurement showing the desorption curve over the temperature. Experiment was carried out by heating the sample from room temperature up to 450 °C in static vacuum (starting value 0.02 bar).

Figure 3.2 shows that all the hydrogen is released by the sample in two steps at 350 °C and 390 °C (after 3.5 hours) which correspond to 8.3 wt. %. $Ca(BH_4)_2$ offers a theoretical hydrogen storage capacity of 11.55 wt. %. However, the reactions reported in section 1.1 indicate that the amount of hydrogen desorbed varies in dependence of the decomposition path. Besides the evolution of hydrogen, the decomposition of $Ca(BH_4)_2$ always leads to the formation of CaH_2 and a boron-compound. Reaction 7 reports the formation of CaH_2 and CaB_6 as decomposition products. This reaction produces 9.6 wt. % hydrogen. Reaction 8, beside the presence of CaH_2, leads to the formation of boron. The amount of hydrogen released in this case is 8.7 wt. %. The value observed after the decomposition of the pure non-milled $Ca(BH_4)_2$ (8.3 wt. %) approaches the lower value reported by reaction 8 (8.7 wt. %) which could be interpreted as a first hint to this reaction. Due to the possible presence of nanocrystalline or amorphous boron compounds among the decomposition products, the combination of X-ray diffraction and ^{11}B MAS-NMR is necessary. In fact, it is well known that XRD is mainly employed to characterise crystalline phases but it fails when adopted to detect nanosized or disordered compounds. Since $^{11}B\{^1H\}$ MAS-NMR does not suffer this limitation it will be used as supportive tool in this study.

3.1.2 Thermal Analysis–Mass Spectrometry

With the purpose of investigating the ongoing series of events taking place during the decomposition of pure non-milled $Ca(BH_4)_2$, differential scanning calorimetry coupled to mass spectrometry measurement was carried out. The sample was measured heating from room temperature up to 550 °C with a constant heating rate of 5 °C min^{-1} in 150 ml min^{-1} argon flow. The results are reported in Figure 3.3.

Figure 3.3 shows three different endothermic reactions. A weak endothermic signal at 165 °C that cannot be appreciated in the Figure due to the graph scale corresponds to the low temperature α- to the high temperature β- phase transformation. Mass spectrometry evidences no hydrogen or diborane gas evolution during this transformation. The first hydrogen desorption step starts around 350 °C (peak temperature 371 °C), followed by the second one in the range of 390-500 °C (450 °C peak temperature). The hydrogen mass signal displays a significant release of hydrogen gas for the aforementioned two endothermic processes. These reaction temperatures perfectly match those observed during the volumetric measurement (Figure 3.2).

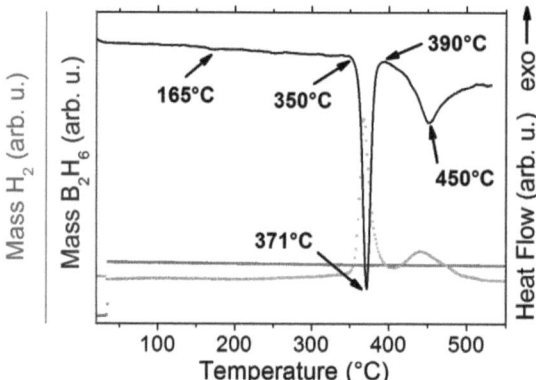

Figure 3.3. DSC (—) and MS of H_2 (□) and of B_2H_6 (□) curves of pure non-milled $Ca(BH_4)_2$ at 150 ml min^{-1} argon flow. The heating rate is 5 °C min^{-1} in 150 ml min^{-1} argon flow.

The concentration of B_2H_6 gas was monitored throughout the experiment. Apart from an increase of its initial value at the beginning of the measurement due to the stabilisation of the spectrometer (effect observed at the same time for O_2, Ar and H_2 as well), no evolution of

B_2H_6 was observed during decomposition of pure non-milled $Ca(BH_4)_2$ in these experimental conditions. This result is in agreement with the study performed by Friedrichs et al.[88]

3.1.3 *In-situ* synchrotron radiation powder X-ray Diffraction

In-situ Synchrotron Radiation Powder X-ray Diffraction was carried out on pure non-milled $Ca(BH_4)_2$ in order to understand the reaction mechanism during the decomposition reaction. The collected sequence of diffraction patterns is reported over the temperature in Figure 3.4. The powder, contained in a sapphire capillary, was heated from room temperature (25 °C) up to 500 °C in static vacuum with a constant heating rate of 5 °C min^{-1}. The measurement was performed at the synchrotron MAX-lab, Lund (Sweden) at the beamline I711.

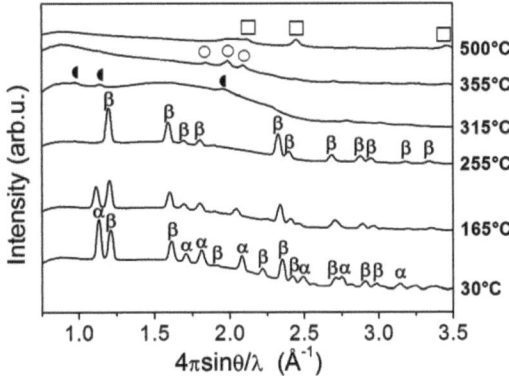

Figure 3.4. SR-PXD patterns of pure non-milled $Ca(BH_4)_2$. The experiment was carried out by heating in vacuum from RT up to 500 °C with 5 °C min^{-1} as heating rate. α-$Ca(BH_4)_2$ (α); β-$Ca(BH_4)_2$ (β); $Ca_3(^{11}BH_4)_3(^{11}BO_3)$ (◆); CaH_2 (○); CaO (□). The measurement was performed at the synchrotron MAX-lab, Lund (Sweden) at the beamline I711.

The starting powder, as shown by the pattern at 30 °C, contains the α- and β-$Ca(BH_4)_2$ phase. Their relative abundance is already reported in section 3.1.1. At 165 °C takes place the low temperature α- to the high temperature β-$Ca(BH_4)_2$ phase transformation. The α-phase continues its conversion into the β-phase until 255 °C where it becomes the only phase visible. Filinchuk et al.[71] reported this phase transformation occurring within the 180-300 °C temperature range. Therefore our findings are in general agreement with the literature. At

255 °C, the Bragg peaks of the β-Ca(BH$_4$)$_2$ phase start to decrease their intensity. Although simultaneous measurement of the variation of pressure during *in-situ* XRD was not performed, this event identifies the beginning of the first hydrogen desorption step which involves only the β-Ca(BH$_4$)$_2$ phase which contains considerable amount of hydrogen. Figure 3.2 and 3.3 illustrated that the first hydrogen desorption step begins at 350 °C. Therefore, between the *in-situ* XRD experiment and both the volumetric and calorimetric analysis there is inaccuracy (± 100 °C) in the temperatures measured. Although this inaccuracy will underestimate the temperatures of the transformations it will not contribute negatively to the understanding of the reaction mechanism. The pattern at 315 °C evidences only the reflections of the δ-Ca(BH$_4$)$_2$ phase. In a previous work[64], this phase was described as another calcium borohydride polymorph but it was recently identified as Ca$_3$(^{11}BD$_4$)$_3$(^{11}BO$_3$).[72] The oxygen could either derive from impurities in the as-synthesised material or from environment (atmosphere, sample holder).[72] At 315 °C, due to the decomposition of β-Ca(BH$_4$)$_2$, we should observe the reflections of the CaH$_2$, CaB$_6$ or boron (amorphous) phase, but none of them shows up in the XRD pattern. The spectrum (315 °C) shows, besides the presence of δ-Ca(BH$_4$)$_2$/Ca$_3$(^{11}BH$_4$)$_3$(^{11}BO$_3$), a broad background within the 0.5-2.5 scattering vector range. This range includes the area where the peaks of the CaH$_2$ and CaB$_6$ phase should appear. A reasonable explanation for the absence of these peaks might be that such a background merges their crystallographic reflections. The diffraction spectrum at 355 °C presents the reflections of the CaH$_2$ phase. Formation of CaH$_2$ indicates the end of the second hydrogen desorption reaction from the δ-Ca(BH$_4$)$_2$/Ca$_3$(^{11}BH$_4$)$_3$(^{11}BO$_3$) phase. The diffraction pattern measured at 500 °C shows the presence of CaO as a side product. The CaO phase originates from the decomposition of Ca$_3$(^{11}BD$_4$)$_3$(^{11}BO$_3$).[72] The pattern at 500 °C does not clearly evidence the peaks of the CaH$_2$ phase anymore. However, at this temperature, it must be still present because its decomposition temperature is ca. 880 °C.[89] Its crystallographic peaks are likely merged with the background.

A study of Riktor *et al*.[64] on Ca(BH$_4$)$_2$, indicated that hydrogen desorption from β-Ca(BH$_4$)$_2$ occurs in the range 330-380 °C, whereas δ-Ca(BH$_4$)$_2$ releases hydrogen at higher temperature, in the range of 380-500 °C. Although Fig. 3.4 shows β-Ca(BH$_4$)$_2$ desorbing hydrogen (first step) in the 255-315 °C temperature range and δ-Ca(BH$_4$)$_2$/Ca$_3$(^{11}BH$_4$)$_3$(^{11}BO$_3$) further desorbing hydrogen (second step) in the 315–355 °C temperature range, Fig. 3.2 and Fig. 3.3 report the hydrogen desorption reactions occurring in the 350-390 °C (first step) and 390-500 °C (second step). These two last ranges are the correct desorption temperature values. As mentioned above, the *in-situ* XRD experiment (Fig. 3.4), is characterised in this specific case, by underestimated desorption temperatures (ca. 100 °C) due to the setup.

No boron-phases (e.g. boron, CaB_6, $CaB_{12}H_{12}$) are visible after hydrogen desorption in Figure 3.4 inferring to their amorphous or nanocrystalline status. To detect those phases $^{11}B\{^1H\}$ MAS-NMR experiments were performed on the (de)hydrogenated samples and they will be presented in section 3.1.5.

3.1.4 The (Re)hydrogenation Reaction

Reversibility represents for hydrogen storage systems a key requirement. The following part of the study is focused on the investigation of the (re)hydrogenation reaction. The products of desorption were (re)absorbed at 350 °C and 130 bar H_2 for 24 hours. The materials were maintained for such a long time at this temperature and pressure because of the commonly known slow absorption reaction kinetics of tetrahydroborates. Under these conditions of high pressures, high temperatures and slow kinetics it is very difficult to monitor accurately absorption curves. Because of their low quality, the obtained curves do not provide any clear information about the quantity of hydrogen (re)absorbed and are thus not reported here. X-ray diffraction measurements are performed on the (re)absorbed material in order to determine whether the (re)absorption reaction was successful. A subsequent desorption measurement, again on the (re)absorbed material, was carried out to determine the amount of hydrogen reversibly absorbed. Figure 3.5 shows the XRD pattern after (re)hydrogenation reaction.

Although the desorbed materials were exposed to a high hydrogen pressure of 130 bar hydrogen and high temperature of 350 °C for 24 hours, no reflection belonging to any $Ca(BH_4)_2$ polymorphs is detectable in Figure 3.5. The Figure still shows the peaks of both the desorbed products (CaH_2 and CaO (side product)). Furthermore, Figure 3.5 does not evidence any crystallographic peak belonging to any boron-phase which is likely amorphous or nanocrystalline.

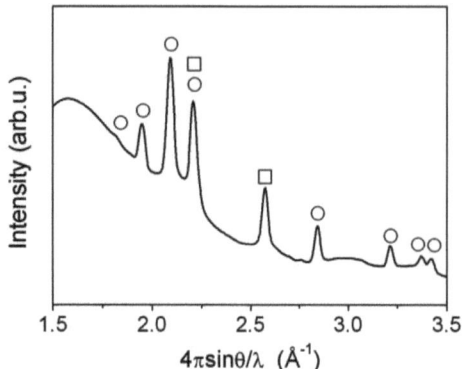

Figure 3.5. SR-PXD pattern of Ca(BH$_4$)$_2$ after (re)absorption reaction at 350 °C and 130 bar H$_2$ for 24 h. CaH$_2$ (O); CaO (□). The measurement was performed at the synchrotron MAX-lab, Lund (Sweden) at the beamline I711.

3.1.5 ^{11}B{^{1}H} Magic Angle Spinning–Nuclear Magnetic Resonance

The desorbed materials were analysed by ^{11}B{^{1}H} MAS-NMR in order to determine which boron phase was formed after hydrogen desorption. The NMR spectrum of the desorbed material is shown in Figure 3.6 together with those of selected reference compounds.

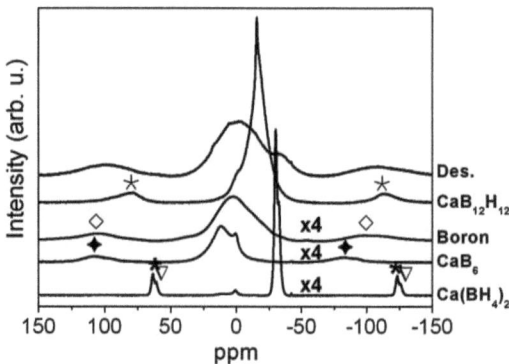

Figure 3.6. ^{11}B{^{1}H} MAS-NMR spectra at room temperature of Ca(BH$_4$)$_2$ desorbed at 450 °C in vacuum (Des.). Ca(BH$_4$)$_2$ purchased by Sigma-Aldrich, CaB$_{12}$H$_{12}$, CaB$_6$ and boron, scale adjusted by ¼. Side bands are indicated by ✶, ▽, ◆, ◊, ★.

Spinning side bands are reported in the Figure as symbols. Commercial Ca(BH$_4$)$_2$ presents two sharp lines at -30 and -32 ppm belonging to the boron atom within the [BH$_4$]$^-$ anion. Since the starting material is composed of the two polymorphs α and β, with different crystal structures (different boron coordination), every peak is related to a different phase. The signal at -30 ppm corresponds to the low temperature phase α (orthorhombic), while the one at -32 ppm belongs to β-Ca(BH$_4$)$_2$ (tetragonal). CaB$_{12}$H$_{12}$ presents a strong signal at -15.4 ppm. This value agrees well with the chemical shift already reported in literature for [B$_{12}$H$_{12}$]$^{2-}$ species (-15.6 ppm).[90] Boron shows a broad signal at ca. +2.5 ppm. The CaB$_6$ spectrum exhibits two lines, at +12 and +0.75 ppm, because of the two different boron sites in its structure.[91] The spectrum of the desorbed material presents two broad peaks at -1.7 and -32 ppm which suggest presence of boron and β-Ca(BH$_4$)$_2$ respectively. A small shoulder at +16 ppm is visible, linked to the formation of small amount of CaB$_6$. The relative content of calcium hexaboride is much less compared to the boron as can be observed in Figure 3.6. The Figure does not evidence the formation of the CaB$_{12}$H$_{12}$ phase. However, it cannot be excluded that a small amount has been formed since its detection is not straightforward due to the broad peak of the boron phase that hides other low intensity peaks over a wide chemical shift range.

3.1.6 Transmission Electron Microscopy

The TEM pictures, the EDX spectrum and the SAED image for the Ca(BH$_4$)$_2$ sample fully desorbed at 450 °C in vacuum are reported in Figure 3.7. The powder was dispersed in THF under argon atmosphere in a glove box and then, a couple of drops were placed dropwise onto a holey carbon supported grid.

At this stage, it is well known that the desorbed Ca(BH$_4$)$_2$ powder contains boron, CaH$_2$ and CaO. Figure 3.7 A and B indicate no clear separation among crystalline (CaH$_2$ and CaO phase) or amorphous areas (boron). The phases appear fairly mixed together. The EDX spectrum (Spec2) evidences Ca signals referred to the CaH$_2$ and CaO phase.

Figure 3.7 A and B show darker and lighter areas. Generally, the difference between the lighter and darker zones is related to the amount of material under analysis. In those parts where the sample is thicker, the transmission of electrons through the specimen is hindered therefore they look darker. Sometimes the contrast is related to differences in the atomic number or density of the material but in our case it is mostly linked to dissimilarities in the amount of sample under analysis.

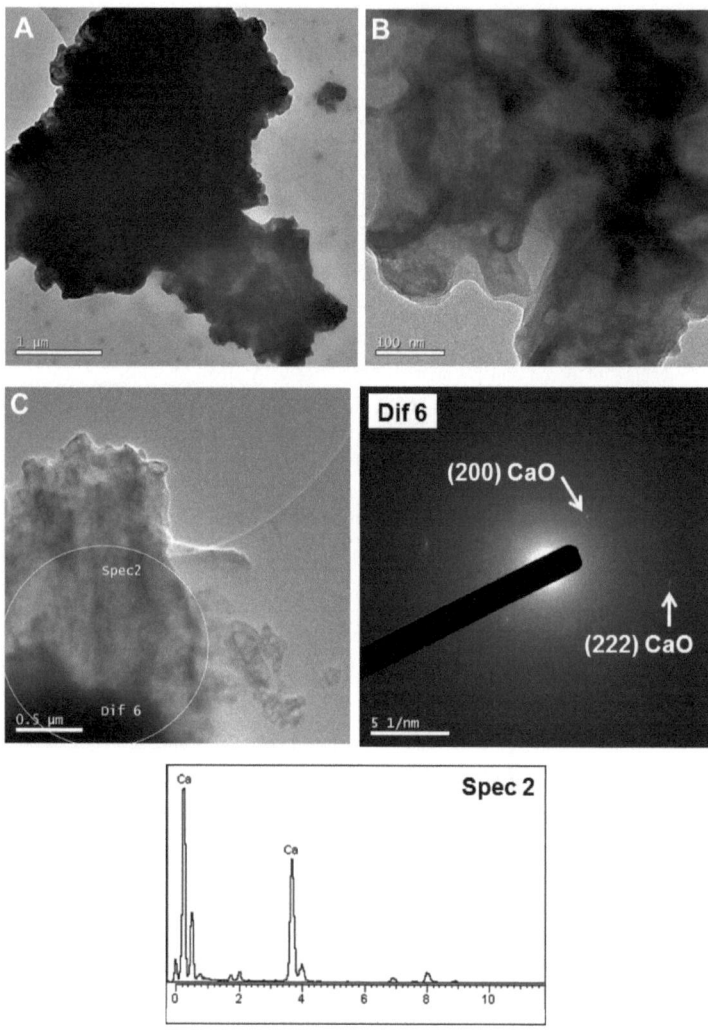

Figure 3.7. TEM image overview (A) of pure non-milled Ca(BH$_4$)$_2$ sample fully desorbed at 450 °C in vacuum. B: detail. C: TEM picture overview. Dif 6: Selected Area Electron Diffraction (SAED) of portion of the material evidenced by a circle in Fig. C. Spec 2: Energy-Dispersive X-ray Spectroscopy (EDX) spectrum of portion of the material evidenced by a circle in Fig. C.

The indexed SAED image (Dif 6) evidences only CaO phase. The average grain size of the material in this area is rather big. Therefore it is not easy to find all the phases within a same area which are aligned to diffract. The SAED image, in this case, shows only one scattering phase (CaO) dispersed in an amorphous background.

Unfortunately a higher resolution of the TEM pictures is not possible because increasing the magnification reduces the contrast and the sample suffers by high sensitivity of the electron beam which causes its decomposition.

3.2 Effect of transition metal fluorides on the sorption properties of $Ca(BH_4)_2$

Section 3.1 has shown the (re)hydrogenation reaction of pure $Ca(BH_4)_2$ to be unsuccessful under the experimental conditions reported in this work. Formation of boron upon decomposition is not beneficial and hinders the reversibility, therefore it should be suppressed. Boron is known to be a stable chemical compound, reluctant to react even in harsh conditions. Such chemical inertness might be due to the strong B-B bond within the elemental boron ($\Delta H_{B(s)-B(g)} = 560$ kJ mol^{-1}).[92]

The addition of additives (metals, oxides, transition metal halides, organic compounds), which decrease the high kinetic barrier, was shown to be essential to promote the reversible hydrogenation reaction of some tetrahydroborates.[25, 93, 94] In particular, transition metal fluorides were considered promising because of the possibility of tuning the thermodynamics of complex hydrides by substituting the H^- with the F^- anion.[50-52, 54, 55] In principle, such a substitution is possible because fluorine and hydrogen have similar atomic radii and exhibit similar valency (-1) in most compounds.[51] The combination of fluorine with transition metal cations takes origin from the catalytic effect of transition metals. In fact, their multivalency could favour electronic exchange reactions with hydrogen molecules, accelerating the gas–solid reaction.

Ma et al.[95] report improved catalytic performances for the TiF_3 doped MgH_2 mixture. By XPS (X-ray Photoelectron Spectroscopy), TiF_3 has generated metastable active Ti–F–Mg species already during milling with MgH_2.[95] This positive effect seems to be related to the different binding state between F and Cl anion.[95] It looks like that F^- is more efficient than Cl^- in tuning the activity of the catalyst.

In case of MgB_2 doped with Ti, TEM investigations revealed a distribution of fine TiB_2 nanoparticles both in the grain boundaries and in the matrix.[96] If the precipitate moves to the grain boundaries it could prevent coarsening phenomena. This effect would stabilise the particle size upon cycling providing faster sorption kinetics. This was already observed by Dornheim et al.[58] in case of MgH_2 doped with transition metal oxides. In that study, the transition metal oxides contribute not only to speed up the surface reaction kinetics but also to reduce and stabilise the MgH_2 crystallite sizes upon cycling at elevated temperatures.

Small amount of transition metal fluorides (NbF_5 and TiF_3 the most effective) evidenced a significant catalytic enhancement of the hydrogenation kinetics of MgH_2.[97, 98] The transition metal fluoride reacts with MgH_2 to produce a hydride-phase or a metal–hydrogen solid solution which forms along the grain boundaries of MgH_2.[97] The hydride layer limits the grain growth of MgH_2 thus preserving its initial catalytic activity.[97] It seems that the phases acting as a catalyst are not the transition metal fluorides, but the hydride phase formed, although a supportive role by MgF_2 (side product) cannot be excluded.[97]

In case of MgH_2 mechanically alloyed with a few % of FeF_3, a fluoride transfer reaction from Fe to Mg and the formation of Fe nanoparticles within the grains of MgH_2 powder was observed. FeF_3-containing MgH_2 powder exhibits an improved H-sorption kinetics (absorption in 200 seconds) at 300 °C.[99]

A treatment of the surface of hydrogen absorbing alloys by an aqueous solution containing F^- anions was shown to lead to a considerable kinetic improvement.[100] Sorption rates are improved because the fluoride layer which is permeable to hydrogen has replaced the oxide layer on the surface. In addition, the mechanical treatment of Mg with inorganic salts (NaF and MgF_2)[101] has shown to promote both the reduction of the crystallites size and to increase its specific area. This approach resulted in faster sorption kinetics in particular during the first hydrogenation cycle. The inorganic salt locates on the Mg surface and spreads along it like a lubricant.[101]

Due to the inability of the desorption products of $Ca(BH_4)_2$ to react together reversibly, the addition of transition-metal fluorides to the tetrahydroborates was performed. Due to the high reactivity of transition-metal fluorides, we consider that they likely react during milling or during desorption in order to lead to the formation of transition metal nanoparticles. The nanoparticles, when homogeneously distributed in the matrix, could favour mass transport, retention of the crystallite size within nanometer order and might act as heterogeneous nucleation agents of compounds like CaB_6 (suppressing boron formation). In addition presence of fluorine-phases on the surface of the tetrahydroborates could accelerate the reaction with hydrogen.

Five different samples were prepared adding 0.05 mol of TiF$_3$ (unknown purity), TiF$_4$ (purity 98%), VF$_3$ (purity 98%), VF$_4$ (purity 95%), NbF$_5$ (purity 99%), purchased from Alfa Aesar, to Ca(BH$_4$)$_2$. A description of the sample preparation is provided in the experimental section 2.2.1.

The characterisation of the sorption properties of the Ca(BH$_4$)$_2$ samples milled with additives will be reported in the following sections by means of volumetric measurements, calorimetric techniques, *in* and *ex-situ* XRD analysis, EXAFS, ^{11}B Magic Angle Spinning-Nuclear Magnetic Resonance and Transmission Electron Microscopy.

3.2.1 The (De)hydrogenation Reaction

The volumetric analysis of the first hydrogen desorption reaction of Ca(BH$_4$)$_2$ milled with additives was performed in a Sievert's type apparatus heating from room temperature (25 °C) up to 450 °C with a constant heating rate of 3 °C min^{-1} under vacuum (0.02 bar the starting pressure value).

X-ray diffraction patterns for all the materials milled with additives are presented in Figure 3.8.

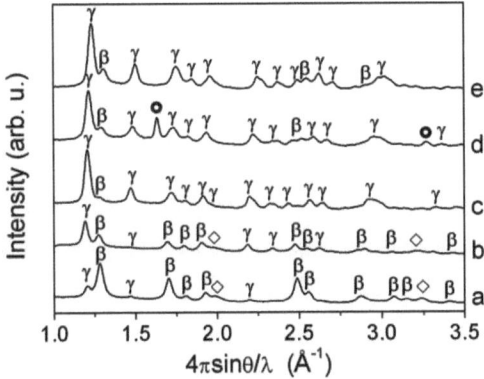

Figure 3.8. XRD of the samples after ball milling: Ca(BH$_4$)$_2$ + TiF$_4$ (a), Ca(BH$_4$)$_2$ + NbF$_5$ (b), Ca(BH$_4$)$_2$ + VF$_3$ (c), Ca(BH$_4$)$_2$ + TiF$_3$ (d), Ca(BH$_4$)$_2$ + VF$_4$ (e). γ-Ca(BH$_4$)$_2$ (γ); β-Ca(BH$_4$)$_2$ (β); CaF$_2$(◇); TiF$_3$(●) (ICSD # 28783). The measurement was performed at the synchrotron MAX-lab, Lund (Sweden) at the beamline I711.

All the samples indicate the presence of γ-Ca(BH$_4$)$_2$ (space group Pbca, orthorhombic phase)[102] and β-Ca(BH$_4$)$_2$ [71] with different abundance. The diffraction patterns (a) and (b), related to the samples milled with TiF$_4$ and NbF$_5$, besides the two polymorphs (γ and β), show reflections of CaF$_2$ (PDF # 35-819). Its presence hints to an irreversible reaction between the TM-fluoride and the borohydride already during the milling process. In contrast, in pattern (d) we can still find peaks of TiF$_3$ (ICSD # 28783), indicating that no reaction has taken place within the vial. Patterns (c) and (e), with VF$_3$ and VF$_4$ respectively, present neither traces of CaF$_2$ nor of the TM-fluorides which suggests a rather fine distribution of the additives.

Figure 3.9 shows the thermovolumetric kinetic curves corresponding to the first desorption reaction of the samples milled with the additives. The desorption curve for pure non-milled Ca(BH$_4$)$_2$ is shown again for comparison purposes.

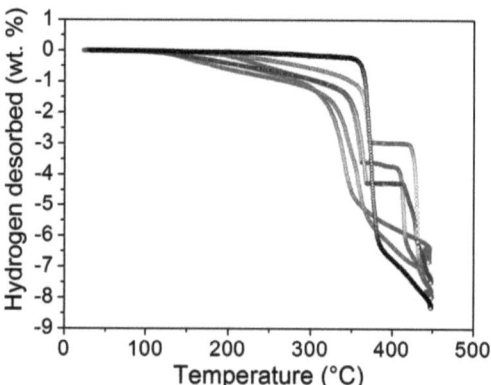

Figure 3.9. Volumetric measurements showing the desorption curves over the temperature. Ca(BH$_4$)$_2$ (black), Ca(BH$_4$)$_2$ + NbF$_5$ (red), Ca(BH$_4$)$_2$ + TiF$_3$ (green), Ca(BH$_4$)$_2$ + TiF$_4$ (dark yellow), Ca(BH$_4$)$_2$ + VF$_3$ (magenta), Ca(BH$_4$)$_2$ + VF$_4$ (blue). Experiments were carried out by heating the samples from room temperature up to 450 °C in static vacuum (starting value 0.02 bar).

Figure 3.9 evidences that the addition of TM-fluorides changes the desorption kinetics: in pure non-milled Ca(BH$_4$)$_2$ the hydrogen starts to be released around 350 °C while in the samples with additives hydrogen release begins already in the range of 125-225 °C. In case of the samples milled with TiF$_3$, VF$_3$ and VF$_4$, the desorption stops for a limited time ($\approx$ 20 minutes) in the range of 360-415 °C and 3-4.5 wt. % desorbed hydrogen. Subsequently, after this incubation, the hydrogen desorption proceeds again until the end of the reaction. In order to collect the materials for XRD and NMR analysis, additional desorption experiments have been performed (with the same experimental conditions) for the same compositions but on different sample batches. The experiments were carried out by heating the samples from room temperature up to 450 °C in static vacuum (starting value 0.02 bar) up to the plateau region. However, as Figure 3.10 shows, for the second set of results (symbol curves), the plateau does not show up for the samples with TiF$_3$, VF$_3$ and VF$_4$.

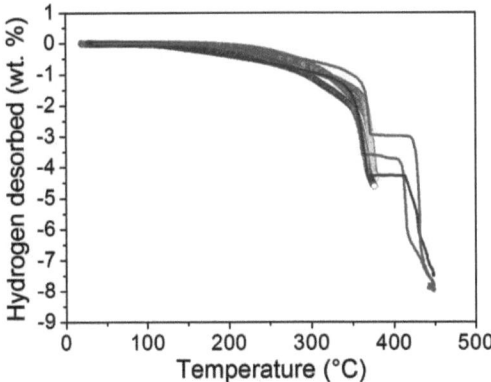

Figure 3.10. Volumetric measurements showing the desorption curves contained in Fig. 3.9 (VF$_3$, VF$_4$ and TiF$_3$) and those corresponding to the experiments performed up to the plateau region (for the same additives). Both experiments were carried out in the same experimental conditions by heating the samples from room temperature up to 450 °C in static vacuum (starting value 0.02 bar). Ca(BH$_4$)$_2$ + TiF$_3$ (green), Ca(BH$_4$)$_2$ + VF$_3$ (magenta), Ca(BH$_4$)$_2$ + VF$_4$ (blue).

Such an incubation period might be caused by a nucleation and growth process of an intermediate calcium borohydride phase as observed in other systems.[93, 103]
This, however, would implicate that by TiF$_3$, VF$_3$ and VF$_4$ the reaction path is altered if compared to pure Ca(BH$_4$)$_2$ where such an incubation period is not observed.

XRD measurements at the plateau, performed for all the materials containing additives, are reported in Figure 3.11.

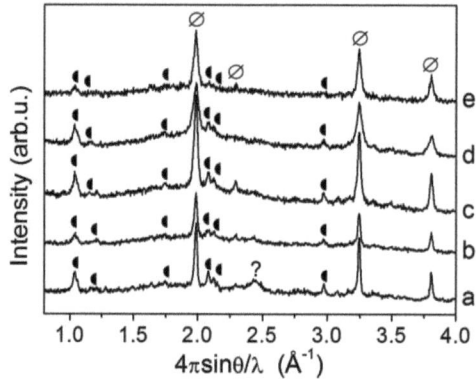

Figure 3.11. XRD patterns of the samples collected at the plateau during the desorption performed up to 450°C in vacuum. Ca(BH$_4$)$_2$ + TiF$_3$ (a); Ca(BH$_4$)$_2$ + VF$_3$ (b); Ca(BH$_4$)$_2$ + VF$_4$ (c); Ca(BH$_4$)$_2$ + NbF$_5$ (d); Ca(BH$_4$)$_2$ + TiF$_4$ (e). Ca$_3$(^{11}BH$_4$)$_3$(^{11}BO$_3$) (◆); unidentified phase (?); CaF$_{2-x}$H$_x$ (∅). The measurements were performed at the Institute for Metallic Materials at the Leibniz Institute for Solid State and Materials Research (Dresden).

Concerning the samples with NbF$_5$ and TiF$_4$, curves (d) and (e) respectively, they were collected stopping the experiment when half of their hydrogen content was desorbed. This can be approximately considered being in the plateau range. In Figure 3.11, for all the samples, two phases are visible: Ca$_3$(^{11}BH$_4$)$_3$(^{11}BO$_3$) and CaF$_{2-x}$H$_x$.[104] The latter is formed by a reaction of CaH$_2$ and CaF$_2$. Only the pattern (a) shows a further peak, indicated by a question mark, not corresponding to any of the phases included in the crystallographic ICSD database. To some extent, although some of the samples show an incubation period during desorption, the X-ray patterns do not evidence relevant differences.

As visible in Figure 3.9, the material doped with NbF$_5$ and TiF$_4$ desorbs at markedly lower temperatures than pristine and doped (TiF$_3$, VF$_3$ and VF$_4$) Ca(BH$_4$)$_2$. These samples (milled with NbF$_5$ and TiF$_4$) do not show any incubation period. This could be due to a better distribution of heterogeneous, amorphous or nanocrystalline, nucleation agents as suggested by Bösenberg *et al.* and Deprez *et al.* for other systems. [93, 103] In case of NbF$_5$ doped

Ca(BH$_4$)$_2$ sample, the improved desorption kinetics, might be linked to a very fine distribution of the transition-metal fluoride additive within the milled powder because of the low melting temperature of NbF$_5$ itself (m. p. 79 °C). In fact, temperatures of ca. 100 °C can be reached within a vial during milling process.[105] This would be more than enough to induce the melting of NbF$_5$. In case of TiF$_4$, the melting temperature is 377 °C which suggests that the transition-metal fluoride unlikely becomes liquid during milling. However, as pointed out at p. 17, its desorption performance seems to be linked to the simultaneous *in-situ* formation and decomposition of Ti(BH$_4$)$_3$. Furthermore, TiF$_4$ might have generated metastable active Ti–F–Ca species during milling with Ca(BH$_4$)$_2$. This behaviour was observed by Ma *et al.*[95] in case of TiF$_3$ doped MgH$_2$ mixture (Ti-F-Mg). This positive effect seems to be related to the different binding state between F and Cl anion.[95] It looks like that F$^-$ is more efficient than Cl$^-$ in tuning the activity of the catalyst.

While the pure non-milled Ca(BH$_4$)$_2$ desorbs 8.3 wt. % hydrogen in two steps at 350 °C and 390 °C (after 3.5 hours) under the applied conditions, the samples with additives desorb less hydrogen because of the formation of some side products like CaF$_2$, which is formed during milling due to an irreversible reaction between the borohydrides and the fluoride additives, and CaO, present among the decomposition products (shown later). Ca(BH$_4$)$_2$ doped with NbF$_5$ and TiF$_4$ start to desorb hydrogen at 120 °C. The reaction speeds up at 300 °C and 315 °C for NbF$_5$ and TiF$_4$ doped samples respectively. After 3.5 hours, Ca(BH$_4$)$_2$ doped with NbF$_5$ and TiF$_4$ desorbs ca. 6.5 wt. % and ca. 7.3 wt. % H$_2$ respectively. Both samples with TiF$_3$ and VF$_3$ desorb ca. 7.8 wt. % H$_2$ whereas the one with VF$_4$ releases ca. 7.5 wt. % H$_2$.

3.2.2 Thermal Analysis

As Figure 3.12 A and B show, there is a clear effect of the investigated fluoride additives on the shape and the onset temperatures of the peaks of the first and the second hydrogen desorption steps.

Curves (f) and (c) in Figure 3.12 A, belonging to the samples milled with NbF$_5$ and TiF$_4$ respectively, show that the first desorption step takes place at significantly lower temperature than in case of the one of the non-milled Ca(BH$_4$)$_2$ as well. In particular, with NbF$_5$, desorption starts below 300 °C. Such a shift in the desorption temperatures is not found in case of the tested vanadium based fluoride additives (curves (d) and (e) in Figure 3.12 A). In this case, the first desorption step is not shifted toward lower temperatures but is even delayed.

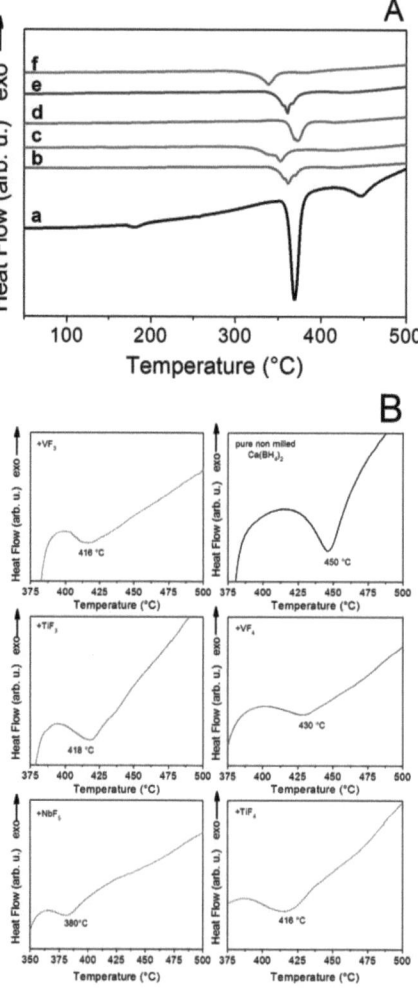

Figure 3.12. A: DSC curves at 150 ml min^{-1} argon flow of Ca(BH$_4$)$_2$ (a); Ca(BH$_4$)$_2$ + TiF$_3$ (b); Ca(BH$_4$)$_2$ + TiF$_4$ (c); Ca(BH$_4$)$_2$ + VF$_3$ (d); Ca(BH$_4$)$_2$ + VF$_4$ (e); Ca(BH$_4$)$_2$ + NbF$_5$ (f). B: detail of the second endothermic event for the DSC curves a-f. The measurements were performed at the Institute for Metallic Materials at the Leibniz Institute for Solid State and Materials Research (Dresden).

The second hydrogen desorption signal shown in Figure 3.12 B for all the milled samples becomes broader compared to that of the non-milled material (black curve Fig. 3.12 B) and shift to lower temperatures. The effect is more pronounced for the sample milled with NbF$_5$ (70 °C lower than pure non-milled Ca(BH$_4$)$_2$).

3.2.3 The (Re)hydrogenation Reaction

The desorption products were subsequently (re)absorbed at 145 bar H$_2$ pressure and 350 °C for ~20 hours. Due to the low quality of the volumetric absorption curves, they will not be reported here. In order to obtain quantitative information about the amount of hydrogen reversibly absorbed, subsequent desorption tests on the (re)absorbed powders were performed. The (re)absorbed materials were analysed by FTIR (Fourier Transform Infrared Spectroscopy) to observe whether the [BH$_4$]$^-$ anion was formed. The advantage of this technique is represented by its rapidity of analysis and effectiveness. Figure 3.13 shows the FTIR spectra of all the materials milled with additives together with pure non-milled Ca(BH$_4$)$_2$ as reference.

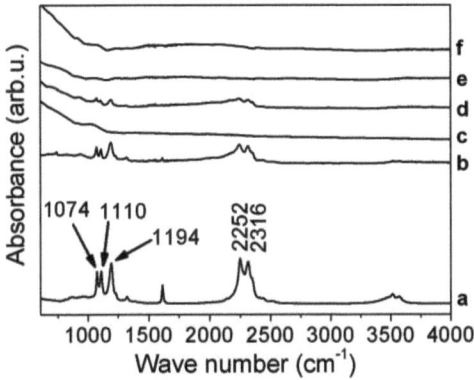

Figure 3.13. FTIR spectra of (re)absorbed powders at 350 °C and 145 bar H$_2$ pressure. Ca(BH$_4$)$_2$ (reference) (a); Ca(BH$_4$)$_2$ + NbF$_5$ (b); Ca(BH$_4$)$_2$ + TiF$_3$ (c); Ca(BH$_4$)$_2$ + TiF$_4$ (d); Ca(BH$_4$)$_2$ + VF$_3$ (e); Ca(BH$_4$)$_2$ + VF$_4$ (f). FTIR measurements were performed at the Institute of Polymer Research at the Helmholtz-Zentrum Geesthacht, Zentrum für Material- und Küstenforschung.

Note that only the NbF$_5$ (spectrum b) and TiF$_4$ (spectrum d) additives were effective in promoting the reversible formation of calcium borohydride. The FTIR signals in the B-H stretching (around 2400 cm^{-1}) and bending regions (around 1180 cm^{-1}) are rather broadened but match those of the Ca(BH$_4$)$_2$ reference indicating reversibility. In contrast, none of the samples with VF$_3$, VF$_4$ or TiF$_3$ additive, which show the incubation period during decomposition, indicate formation of B-H bonds. Other signals at 1615 and around 3500 cm^{-1} are related to water adsorbed at the surface of the powder because the experiments were performed in air.

Although FTIR is a quick and good tool to reveal the presence of the [BH$_4$]$^-$ anion after (re)absorption reaction, it lacks information concerning all the others crystallographic phases present within the system. Therefore, X-ray diffraction was performed on the (re)absorbed samples doped with NbF$_5$ and TiF$_4$ because they were the only materials showing reversible formation of Ca(BH$_4$)$_2$. The XRD patterns for both the samples are reported in Figure 3.14.

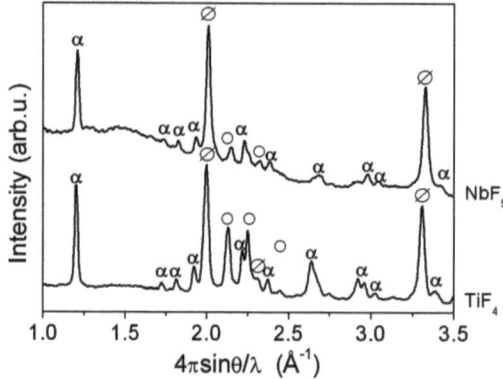

Figure 3.14. XRD of Ca(BH$_4$)$_2$ doped with NbF$_5$ and TiF$_4$ after (re)absorption at 350 °C and 145 bar H$_2$ pressure for ~20 hours. α-Ca(BH$_4$)$_2$ (α); CaF$_{2-x}$H$_x$ (Ø); CaH$_2$ (O). The measurements were performed at the synchrotron MAX-lab, Lund (Sweden) at the beamline I711.

Figure 3.14 displays, for both the patterns, the reflections of the α-Ca(BH$_4$)$_2$ phase. Hence, partial reversible formation was achieved. The (de)hydrogenated products CaH$_2$ and CaF$_{2-x}$H$_x$ are also visible. Note that the starting doped Ca(BH$_4$)$_2$ powder contained both the γ- and β-modification whereas the (re)hydrogenated material shows only the α- modification. This is related to the higher stability of the α- (up to 550 K) in respect to the β- polymorph (stable T>550 K). The γ-phase, instead, is not energetically favoured at any temperature.[106] Furthermore, the simultaneous presence of CaH_2 and $CaF_{2-x}H_x$ phase can be explained by the reversible reactions (1) and (2) reported below. These reactions proceed in dependence of the different atmosphere:

(1) $CaH_2 + CaF_2 \rightarrow CaF_{2-x}H_x$ (vacuum)

(2) $CaF_{2-x}H_x \rightarrow CaH_2 + CaF_2$ (H_2 pressure)

Calculation of the relative phase abundance (by Rietveld method) for the XRD patterns reported in Figure 3.14 is not useful. The presence of amorphous or nanocrystalline phases (e.g. $CaB_{12}H_{12}$ or boron) which cannot be detected by XRD would contribute to overestimate the calculated values. In order to obtain quantitative information about the amount of hydrogen reversibly absorbed, desorption tests on the (re)absorbed powders were performed.

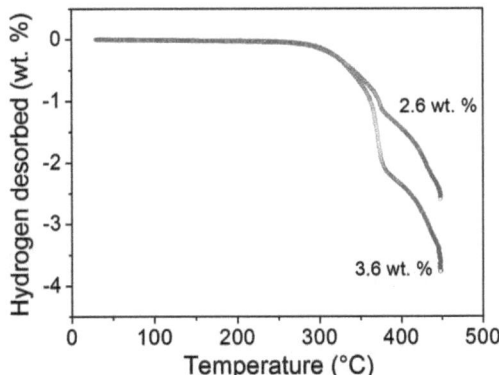

Figure 3.15. Desorption curves after (re)absorption at 350 °C and 145 bar H_2 pressure: $Ca(BH_4)_2$ + TiF_4 (dark yellow), $Ca(BH_4)_2$ + NbF_5 (red). Experiments were carried out by heating the samples from room temperature up to 450 °C in static vacuum (starting value 0.02 bar).

Figure 3.15 shows the thermovolumetric measurements related to the second hydrogen desorption reaction of the samples milled with TiF$_4$ and NbF$_5$. As shown in the Figure, the sample with TiF$_4$ desorbed 2.6 wt. % whereas the sample with NbF$_5$ desorbed 3.6 wt. % hydrogen, corresponding to 35 % and 55 % reversibility, respectively. Figure 3.15 also shows distinct changes of the slope for both the desorption curves, typical of multistep desorption processes.

3.2.4 Ca(BH$_4$)$_2$ + NbF$_5$: *in-situ* Synchrotron Radiation Powder X-ray Diffraction

In-situ SR-PXD was employed to obtain a comprehensive understanding of the sequence of reactions occurring during hydrogen desorption for the samples doped with NbF$_5$ and TiF$_4$. As already shown in the previous sections (3.2.3), these two additives are the only effective in promoting the reversible formation of Ca(BH$_4$)$_2$ under the applied conditions. Therefore a detailed investigation of their reaction mechanisms is necessary. The SR-PXD patterns collected at selected temperatures are reported in Figure 3.16 for the material milled with NbF$_5$.

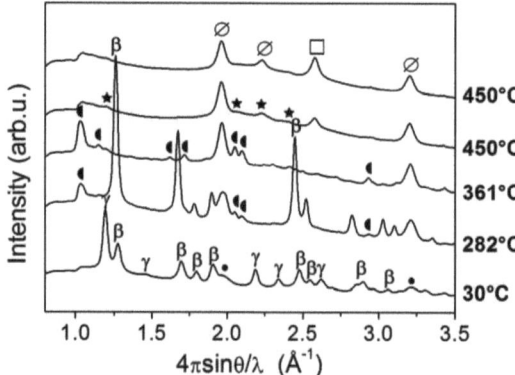

Figure 3.16. SR-PXD patterns of Ca(BH$_4$)$_2$ milled with NbF$_5$. The experiment was carried out by heating the sample in vacuum from RT up to 450 °C with 5 °C min^{-1} as heating rate and 10 minutes isotherm (at 450 °C). γ-Ca(BH$_4$)$_2$ (γ); β-Ca(BH$_4$)$_2$ (β); CaF$_2$ (•) ; Ca$_3$(^{11}BH$_4$)$_3$(^{11}BO$_3$) (◀); unidentified phase (★); CaO (□); CaF$_{2-x}$H$_x$ (∅). The measurement was performed at the synchrotron MAX-lab, Lund (Sweden) at the beamline I711.

Desorption reactions were studied under static vacuum by heating from room temperature up to 450 °C. Rietveld refinement of the pattern collected at 30 °C indicates that the initial powder is composed of 72 wt. % (± 5 error) low temperature polymorph γ-Ca(BH$_4$)$_2$, 24 wt. % (± 5 error) high temperature β-Ca(BH$_4$)$_2$ and 4 wt. % (± 5 error) CaF$_2$ (PDF # 35-819). The latter one must have been formed by an irreversible reaction between Ca(BH$_4$)$_2$ and NbF$_5$. This might partly explain why the total amount of hydrogen desorbed from the samples with additives is lower than in case of the pure calcium borohydride. The pattern measured at 282 °C shows that there are no traces of γ-phase anymore, since it is already transformed into β-phase. At this temperature the peaks of the Ca$_3$(^{11}BD$_4$)$_3$(^{11}BO$_3$) (previously called δ-Ca(BH$_4$)$_2$) phase appear. The work of Riktor et al.[72] reported that hydrogen desorption from β-Ca(BH$_4$)$_2$ occurs in the range of 330 °C to 380 °C, whereas Ca$_3$(^{11}BD$_4$)$_3$(^{11}BO$_3$) releases hydrogen at higher temperature, in the range of 380-500 °C. Based on results reported in Fig. 3.9, 3.12 A-B and the above mentioned SR-PXD patterns, referred to the first desorption reaction, the addition of TM-fluorides to calcium borohydride, modifies its desorption kinetics and the thermal events are shifted towards lower temperatures.

As can be seen in Figure 3.16, the XRD pattern at 282 °C, besides CaF$_2$, evidences the reflections of both the β-Ca(BH$_4$)$_2$ and of the Ca$_3$(^{11}BD$_4$)$_3$(^{11}BO$_3$) phase. The pattern at 361 °C does not show peaks belonging to the β-phase anymore instead reflections corresponding to the CaF$_{2-x}$H$_x$ phase are visible. CaH$_2$ should be formed at this temperature as a consequence of the first desorption event, however, our measurements do not evidence its peaks up to 361 °C. Therefore CaH$_2$, produced from the decomposition of β-Ca(BH$_4$)$_2$, has reacted with the CaF$_2$, already present in the starting mixture, forming CaF$_{2-x}$H$_x$ (reaction 1 and 2, section 3.2.3). This result confirms that the first hydrogen desorption step from β-Ca(BH$_4$)$_2$ takes place between 282 °C and 361 °C . In this range, most of the hydrogen is desorbed with some residual being present inside the Ca$_3$(^{11}BH$_4$)$_3$(^{11}BO$_3$). A comparison between the patterns at 282 °C and at 361 °C shows that this borohydride borate phase fraction grows continuously up to 361 °C, as can be inferred from the increase of the intensity of its diffraction peaks. At 450 °C, the reflections of the Ca$_3$(^{11}BH$_4$)$_3$(^{11}BO$_3$) phase are not visible anymore and CaO and an unidentified phase forms. In the work of Riktor et al.[64] the only Ca$_3$(^{11}BH$_4$)$_3$(^{11}BO$_3$) (or δ-Ca(BH$_4$)$_2$) phase was observed to release hydrogen in the second step (380-500 °C). In the same work [64], the unidentified phase was formed later, at 500 °C, and no gas evolution from it was determined. Hence, disappearance of the peaks corresponding to Ca$_3$(^{11}BH$_4$)$_3$(^{11}BO$_3$) (first pattern at 450 °C) identifies the end of the second hydrogen desorption reaction. Therefore residual hydrogen is released in the 361-450 °C temperature range. At 450 °C, presence of CaO reflections is observed. CaO results from the decomposition of $Ca_3(^{11}BH_4)_3(^{11}BO_3)$ the unidentified phase decomposes after a short time. Later on, the last recorded XRD pattern at 450 °C shows an increase of the peak intensity at the scattering vector value of 2.23 (Å^{-1}). This peak corresponds to the (200) peak of the $CaF_{2-x}H_x$ phase.[104] The ICSD database matches those peaks with an exact stoichiometry $CaF_{0.76}H_{1.24}$, however, the current experimental results are not conclusive concerning this point. Another observation is the increase of the diffraction peak intensity of CaO. The increase of the intensity of the CaO peaks takes place after decomposition of the unidentified phase which probably contains calcium oxide. This would be reasonable since this unidentified phase forms after decomposition of $Ca_3(^{11}BH_4)_3(^{11}BO_3)$ which obviously contains oxygen.

3.2.5 Ca(BH$_4$)$_2$ + TiF$_4$: *in-situ* Synchrotron Radiation Powder X-ray Diffraction

SR-PXD patterns for selected temperatures are reported in Figure 3.17 for the material milled with TiF$_4$.

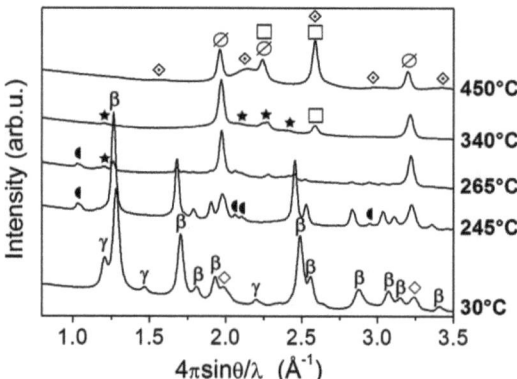

Figure 3.17. SR-PXD patterns of Ca(BH$_4$)$_2$ milled with TiF$_4$. The experiment was carried out by heating the sample in vacuum from RT up to 450 °C with 5 °C min^{-1} as heating rate and 10 minutes isotherm (at 450 °C). γ-Ca(BH$_4$)$_2$ (γ); β-Ca(BH$_4$)$_2$ (β); CaF$_2$ (◇); Ca$_3$(^{11}BH$_4$)$_3$(^{11}BO$_3$) (◉); unidentified phase (★); CaO (□); CaF$_{2-x}$H$_x$ (∅); CaB$_6$ (◈). The measurement was performed at the synchrotron MAX-lab, Lund (Sweden) at the beamline I711.

Rietveld refinement of the diffractogram at 30 °C indicates that the starting material is composed of 14 wt. % (± 5 error) γ-Ca(BH$_4$)$_2$, 81 wt. % (± 5 error) β-Ca(BH$_4$)$_2$ and 5 wt. % (± 5 error) CaF$_2$ (PDF # 35-819). Formation of CaF$_2$, as reported in section 3.2.4, again causes a reduction of the overall hydrogen capacity of the material.

The pattern measured at 245 °C shows the peaks of Ca$_3$(^{11}BH$_4$)$_3$(^{11}BO$_3$), but does not evidence any reflection of γ-Ca(BH$_4$)$_2$ which has totally transformed into β-Ca(BH$_4$)$_2$ at this temperature. The pattern at 265 °C does not show peaks belonging to the β-phase anymore instead reflections corresponding to the CaF$_{2-x}$H$_x$ phase are visible. CaH$_2$ should be formed at this temperature as a consequence of the first desorption event, however, our measurements do not evidence its peaks. As already observed for the Ca(BH$_4$)$_2$ + NbF$_5$ sample (Figure 3.16), CaH$_2$, produced from the decomposition of β-Ca(BH$_4$)$_2$, has reacted with the CaF$_2$, already present in the starting mixture, forming CaF$_{2-x}$H$_x$ (reaction 1 and 2, section 3.2.3). Such a non-stoichiometric Ca-F-H phase explains the absence of CaH$_2$ reflections. This result also confirms that the first hydrogen desorption step from β-Ca(BH$_4$)$_2$ takes place between 245 °C and 265 °C. In this range, most of the hydrogen is desorbed with some residual being present inside the Ca$_3$(^{11}BH$_4$)$_3$(^{11}BO$_3$). At 265 °C an unidentified phase is visible. This phase is exactly the same which was observed in the diffraction pattern of Ca(BH$_4$)$_2$ + NbF$_5$ (Figure 3.16). A comparison between the pattern at 245 °C and at 265 °C shows that the Ca$_3$(^{11}BH$_4$)$_3$(^{11}BO$_3$) phase, in this range of temperature, decrease its fraction as can be inferred by the decrease of the intensity of its main peak, at the scattering vector value of 1.03 (Å^{-1}). This is explained by the formation of the unidentified phase which, as already observed in Figure 3.16, forms after decomposition of Ca$_3$(^{11}BH$_4$)$_3$(^{11}BO$_3$). At 340 °C, the reflections of the Ca$_3$(^{11}BH$_4$)$_3$(^{11}BO$_3$) phase are not visible anymore. As already reported in section 3.2.4, Riktor et al.[64] observed that only the Ca$_3$(^{11}BH$_4$)$_3$(^{11}BO$_3$) (or δ-Ca(BH$_4$)$_2$) phase releases hydrogen in the second step (380-500 °C). In the same work [64], no gas evolution from the unidentified phase was detected because its formation occurred later (at 500 °C). Hence, disappearance of the peaks corresponding to Ca$_3$(^{11}BH$_4$)$_3$(^{11}BO$_3$) (first pattern at 340 °C) identifies the end of the second hydrogen desorption reaction. Therefore, residual hydrogen is released in the 265-340 °C temperature range. At 340 °C, the unidentified phase is still present and more of its reflections are now visible. At this temperature, also the crystallographic peak of the CaO phase is present. Presence of CaO confirms what already observed for the Ca(BH$_4$)$_2$ + NbF$_5$ sample (Figure 3.16): upon heating, decomposition of Ca$_3$(^{11}BH$_4$)$_3$(^{11}BO$_3$)) leads to CaO formation. The unidentified phase decomposes at 450 °C. At this temperature the peak at the scattering vector value of 2.24 (Å^{-1}) shows higher intensity. This peak corresponds to the (200) peak of the $CaF_{2-x}H_x$ phase.[104] After decomposition of $Ca_3(^{11}BH_4)_3(^{11}BO_3)$, no traces of CaH_2 were detected. Presence of Ca-F-H explains the absence of CaH_2 reflections (reaction 1 and 2, section 3.2.3). Another observation in the pattern at 450 °C is the increase of the diffraction peak intensity of CaO compared to the diffractogram at 340 °C. As already reported in section 3.2.4, the increase of intensity of CaO peaks takes place after decomposition of the unidentified phase which obviously contains oxygen. Note that the pattern at 450 °C evidences also reflections of CaB_6 which was not observed before in both pure non-milled (Fig. 3.4) and NbF_5 doped $Ca(BH_4)_2$ material (Fig. 3.16).

3.2.6 $Ca(BH_4)_2$ + NbF_5: X-ray Absorption Near Edge Structure

The addition of NbF_5 to the $Ca(BH_4)_2$ material has demonstrated to be necessary to promote its reversible formation. It was reported in this study that the addition of NbF_5 is more effective than that of TiF_4. Unfortunately, both the reaction mechanism and the influence of the additive are not understood yet. The nature of the Nb-phase as well as its oxidation state during sorption reactions was determined by XANES (X-ray Absorption Near Edge Structure). As already reported in literature by Friedrichs *et al.*[107-109], Deprez *et al.*[103, 110] and Bösenberg *et al.*[111, 112], X-ray based techniques represented a powerful tool to understand the role of the additives in both MgH_2 and $LiBH_4$ + MgH_2 RHC system. XANES data measured at the Nb K-edge (18986 eV) are shown in Figure 3.18. The Figure includes the spectrum of the material after first desorption, after subsequent (re)absorption and after second desorption reaction. The results for the milled sample could not be reported here due to the low quality of the curve.

A direct comparison between the curve of the desorbed material and that of the pure NbF_5 additive indicates that a reaction, involving the additive itself and the $Ca(BH_4)_2$, has occurred during the desorption process. This was already presented by the volumetric curve in Figure 3.9 which evidenced a low temperature reaction in the temperature range of 125-225 °C. Furthermore, the curves of the material after (re)absorption and after second hydrogen desorption indicate no further change in the oxidation state of the Nb over cycling. The shift to lower energy of the Nb K-edge in the desorbed sample in respect to the NbF_5 additive indicates that the oxidation state of the Nb species reduces irreversibly. The irreversible reduction is confirmed observing the curves for the material after (re)absorption and second desorption which are comparable to that after first (de)hydrogenation.

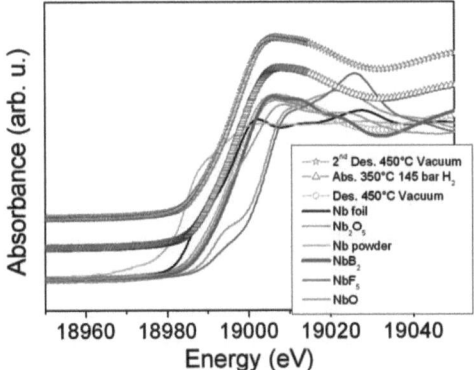

Figure 3.18. XANES data at the Nb K-edge for $Ca(BH_4)_2$ + NbF_5 (with 5 mol % NbF_5) material after first desorption, after subsequent (re)absorption and after second desorption. The measurements were performed at the synchrotron Hasylab, DESY (Hamburg), at the beamline C.

Figure 3.18 shows that the curves belonging to the desorbed/(re)absorbed materials fairly match that of the pure NbB_2, measured as a reference compound. This would indicate a reduction of the oxidation state for the Nb species from (V) to (II). Furthermore, the similar profile for the curves belonging to the desorbed/(re)absorbed materials and for the NbB_2 along the EXAFS (Extended X-ray Absorption Fine Structure) region (at higher energies respect to the Nb absorption K-edge) would suggest a Nb-B bond in the first coordination shell. This assumption needs to be confirmed by the Fourier Transform analysis of the EXAFS data. Note that *ex-situ* and *in-situ* XRD, in Figure 3.8, 3.11, 3.14 and 3.16, do not evidence any trace of NbB_2 phase. As already reported in literature [111], the dimensions of such boride particles can be in the range of a few nanometers only and hence out of the detection range of X-ray diffraction techniques. As visible in Figure 3.17, in the pattern at 450 °C, the CaB_6 phase has an amorphous-like profile. It is therefore necessary to adopt Transmission Electron Microscopy (TEM) in order to observe and determine the size of such nanoparticles. TEM analysis will be reported later. However, the formation of transition metal boride nanoparticles upon hydrogen desorption reaction, in the case of doped systems, is more than a certainty and it would be consistent with several works recently reported in literature.[103, 111, 112]

As presented in this section for the NbF$_5$ additive, the formation of a transition metal boride is observed. This result is consistent with the works recently reported by Deprez et al.[103], Bösenberg et al.[111, 112] and Ngene et al.[113].

3.2.7 Ca(BH$_4$)$_2$ + TiF$_4$: X-ray Absorption Near Edge Structure

The effect that the addition of TiF$_4$, as well as NbF$_5$, represents for the reversibility of Ca(BH$_4$)$_2$ is noteworthy. For the understanding of the reaction mechanism it is necessary to know the nature of the Ti phase formed upon the hydrogen desorption/(re)absorption process. XRD cannot help any further concerning this point because, generally, the dimensions of these phases are within the nanometer range. As indicated previously, XANES represents a powerful tool on this regard because it allows determining the oxidation state of the transition metals.

XANES analysis at the Ti K-edge (4966 eV), including the curve for the ball milled and desorbed material, are reported in Figure 3.19.

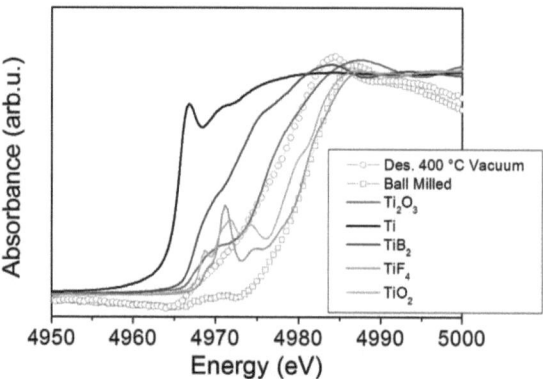

Figure 3.19. XANES data at the Ti K-edge for the milled Ca(BH$_4$)$_2$ + TiF$_4$ (with 5 mol % TiF$_4$) material after ball milling and first desorption. The measurements were performed at the synchrotron Hasylab, DESY (Hamburg), at the beamline A1.

A comparison, in the pre-edge region of the spectra, among the curve of the ball milled material and of the TiF$_4$, TiO$_2$ and Ti$_2$O$_3$, suggests that a reaction, involving TiF$_4$ and Ca(BH$_4$)$_2$, has taken place within the vial already during the milling process. In fact, the pre-edge at 4971 eV for the TiF$_4$ curve has disappeared in the case of the milled sample. The reaction during the milling process leads to the formation of CaF$_2$ as showed by XRD in Figure 3.8 (a).

The Ti absorption K-edge, in the case of the milled sample, matches the value observed for TiO$_2$, Ti$_2$O$_3$ and TiF$_4$. Presence of TiF$_4$ after milling is unexpected since XRDs, in Figure 3.8 (a), beside γ-Ca(BH$_4$)$_2$ and β-Ca(BH$_4$)$_2$, show presence of CaF$_2$ which should be caused by a reaction between Ca(BH$_4$)$_2$ and TiF$_4$. However, part of TiF$_4$ could still be present, due to the milling, in nanometer-size which would make it undetectable by X-rays. In addition, within the rising edge region (at higher energies compared to the pre-edge region) the curve of the milled sample overlaps that of pure TiF$_4$.

Given the formation enthalpy data and considered the high sensitivity of alkali tetrahydroborates to moisture, formation of Ti$_2$O$_3$ seems to be likely.[27] In addition, Riktor et al.[72] reported the starting Ca(BH$_4$)$_2$ material containing already solvent traces or oxygen-containing impurities. However, within the rising edge region, the curve of the milled sample is closer to that of pure TiO$_2$ than that of Ti$_2$O$_3$. Experiments performed by Buslaev et al.[114] would exclude TiO$_2$ as a product of the hydrolysis reaction of TiF$_4$. The combination of the observations reported for the milled sample, pre-edge region, at the rising edge region and at the absorption edge allows to conclude that Ti has an oxidation state of IV (TiO$_2$ or TiF$_4$).

The desorbed sample shows a reduction in the oxidation state of Ti. The Ti absorption K-edge, in the case of the desorbed sample, matches the value observed for TiB$_2$. Within the rising edge region the curve of the desorbed sample fits that of pure Ti$_2$O$_3$. The combination of these observations allows concluding either that Ti has an oxidation state between III (Ti$_2$O$_3$) and II (TiB$_2$) or a mixture of both different chemical states. The same behaviour was already reported by Deprez et al.[103] in case of Ti-isopropoxide doped LiBH$_4$-MgH$_2$ RHC. Unfortunately, the results for the (re)absorbed material could not be reported here due to the low quality of the curve.

3.2.8 ^{11}B{^{1}H} Magic Angle Spinning–Nuclear Magnetic Resonance

SR-PXD experiments (Figure 3.16 and 3.17) show an identical decomposition pathway for the samples milled with NbF$_5$ and TiF$_4$ but still, in the former sample, it is not clear whether CaB$_6$ is present as decomposition product. Reactions 7 and 8, presented in section 1.1, indicate that the decomposition of calcium borohydride ends in boron or CaB$_6$, but X-ray diffraction does not represent the proper tool to detect fine nano-crystalline and/or amorphous compounds. Therefore, ^{11}B{^{1}H} MAS-NMR was carried out to identify the nature of the final B-containing compounds.

^{11}B{^{1}H} MAS-NMR spectra are shown in Figure 3.20. ^{11}B (I=3/2) NMR was collected at room temperature on the desorbed samples. Commercial Ca(BH$_4$)$_2$ by Sigma-Aldrich, CaB$_6$, elemental boron and CaB$_{12}$H$_{12}$ are included as references.

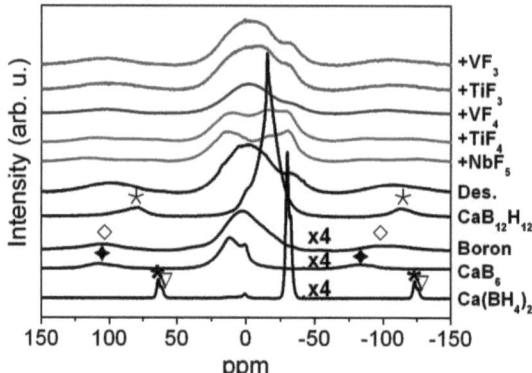

Figure 3.20. ^{11}B{^{1}H} MAS-NMR spectra at room temperature. Ca(BH$_4$)$_2$ purchased from Sigma-Aldrich and CaB$_{12}$H$_{12}$. Ca(BH$_4$)$_2$, CaB$_6$ and boron, scale adjusted by ¼. Side bands are indicated by ✶, ▽, ◆, ◊, ★.

As pointed out in section 3.1.5, the spectrum corresponding to Ca(BH$_4$)$_2$ desorbed at 450 °C in vacuum (Des.), indicates the presence of a broad signal at ~-1.7 ppm characteristic of elemental boron and another less intense, at -32 ppm, corresponding to residual not reacted β-Ca(BH$_4$)$_2$. Broadening in NMR signals is typical of disordered materials that isotropically distribute their chemical shifts.[90] In addition, a small shoulder at +16 ppm is visible, linked to the formation of small amount of CaB$_6$.

The spectrum of desorbed Ca(BH$_4$)$_2$ + NbF$_5$ (red spectrum) shows a series of three signals at ~+16 ppm, ~-16 ppm and ~-30 ppm. The peak at +16 ppm corresponds to CaB$_6$ and the one at -30 ppm evidences residual β-Ca(BH$_4$)$_2$. The identification of the peak at ~-16 ppm is not straightforward. The ^{11}B{^{1}H} MAS-NMR spectrum for the as-synthesised CaB$_{12}$H$_{12}$, reported as reference in Figure 3.6 and 3.20, results in a strong signal at -15.4 ppm.[90, 115] This value agrees well with the chemical shift observed in literature for [B$_{12}$H$_{12}$]$^{2-}$ species (-15.6

ppm). By ^{11}B MAS-NMR, Hwang et al.[90] confirmed the formation of $[B_{12}H_{12}]^{2-}$ anion as intermediate compound during desorption of LiBH$_4$ by the detection of a broad signal at -12 ppm. This value might shift in dependence either of the cation associated to the $[B_{12}H_{12}]^{2-}$ anion and/or the disordered nature of the phase structure.[90]

The Ca(BH$_4$)$_2$ + TiF$_4$ sample desorbed at 450 °C in vacuum (dark yellow spectrum) shows the same series of three broad signals similar to the sample milled with NbF$_5$ (red spectrum), evidencing CaB$_6$, residual β-Ca(BH$_4$)$_2$ and CaB$_{12}$H$_{12}$. Their similar decomposition products reflect their equivalent followed decomposition paths (Fig. 3.16 and 3.17).

The ^{11}B{^{1}H} MAS-NMR spectrum of Ca(BH$_4$)$_2$ + VF$_4$ shows a profile comparable to that of pure desorbed calcium borohydride. The pattern indicates the presence of elemental boron and residual β-Ca(BH$_4$)$_2$ as can be observed by the shoulder at -34 ppm. This value is slightly shifted compared to that observed in other samples (-30 ppm).

As seen in Figure 3.13 and 3.14, under the applied conditions, TiF$_4$ and NbF$_5$ were the only additives capable to promote the reversible formation of calcium borohydride, even though partial. ^{11}B{^{1}H} MAS-NMR shows that they lead to the same boron phases as decomposition products: CaB$_6$ and CaB$_{12}$H$_{12}$. In contrast, elemental boron was found as decomposition product in the samples milled with VF$_4$, TiF$_3$ and VF$_3$. These three TM-fluoride additives do not have any positive influence on the reversible formation of calcium borohydride under the conditions applied in this study.

3.2.9 Transmission Electron Microscopy

TEM microscopy of the desorbed Ca(BH$_4$)$_2$ samples doped with NbF$_5$ and TiF$_4$ was carried out to confirm the presence of the transition metal boride (NbB$_2$ and TiB$_2$) or Ti$_2$O$_3$ nanoparticles in the desorbed/(re)absorbed materials.

The TEM pictures for the desorbed sample doped with NbF$_5$ are reported in Figure 3.21. At first, the detection of the nanoparticles was complex. The powder decomposes within a few seconds due to the influence of the electron beam. Moreover, whenever the magnification was increased to obtain a better picture the contrast decreased significantly. In order to bypass these issues, another approach was undertaken. Since the nanoparticles are crystalline, the investigation in dark field mode could provide pictures with a better contrast.

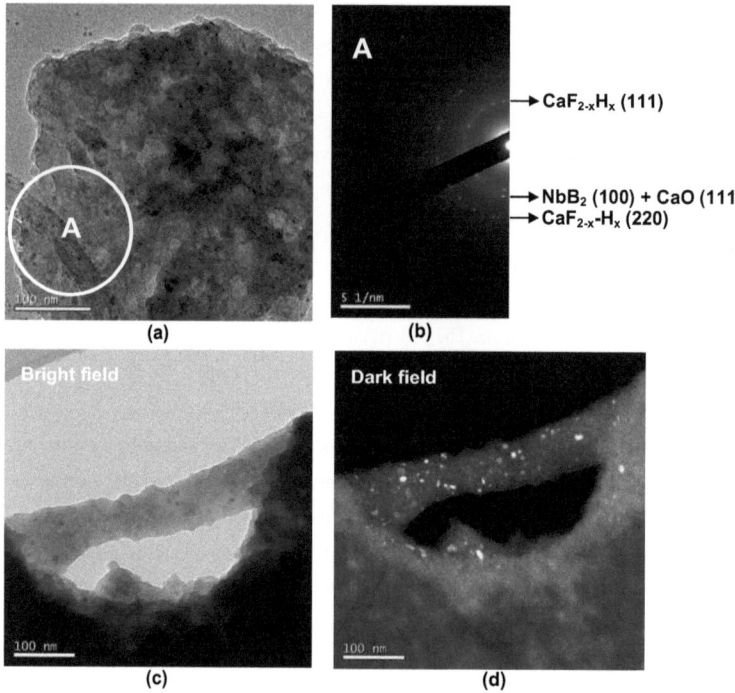

Figure 3.21. TEM images (bright and dark field) together with SAED analysis of the Ca(BH$_4$)$_2$ sample doped with NbF$_5$ fully desorbed at 450 °C in vacuum.

Figure 3.21 (a) shows dark spots dispersed within the matrix of the desorbed materials, corresponding to the NbB$_2$ nanoparticles. This is confirmed by Selected Area Electron Diffraction (SAED), in Figure 3.21 (b), performed in the area indicated as A in Figure 3.21 (a). The diffraction analysis reveals presence of CaF$_{2-x}$H$_x$ phase, CaO and NbB$_2$ (Figure 3.21 (b)). CaB$_{12}$H$_{12}$ cannot be detected due to its amorphous status. Pictures (c) and (d), in bright and dark field mode respectively, correspond to another portion of the sample. Picture (d), in particular, due to the dark field mode (see section 2.10) highlights the nanoparticles within the desorbed materials as bright spots. Although it was not possible to perform electron diffraction in this spot, considering the results reported in Figure 3.21 (a) and (b) and the homogeneity of the sample, we can reasonably assume the bright spots being NbB$_2$ nanoparticles. The average size of the NbB$_2$ nanoparticles is around 10 nm. The lighter grey zone in Figure (d) which embeds the nanoparticles is representative of amorphous phases (e.g. $CaB_{12}H_{12}$).

We can exclude that the bright particles correspond to $CaF_{2-x}H_x$ or CaO because, as reported in the *in-situ* XRD pattern after decomposition (Fig. 3.16), they exist in much bigger size. Instead, no transition metal boride is visible in the same XRD pattern (Fig. 3.16) implying their existence in nanometer size.

The TEM pictures for the desorbed sample doped with TiF_4 are reported in Figure 3.22.

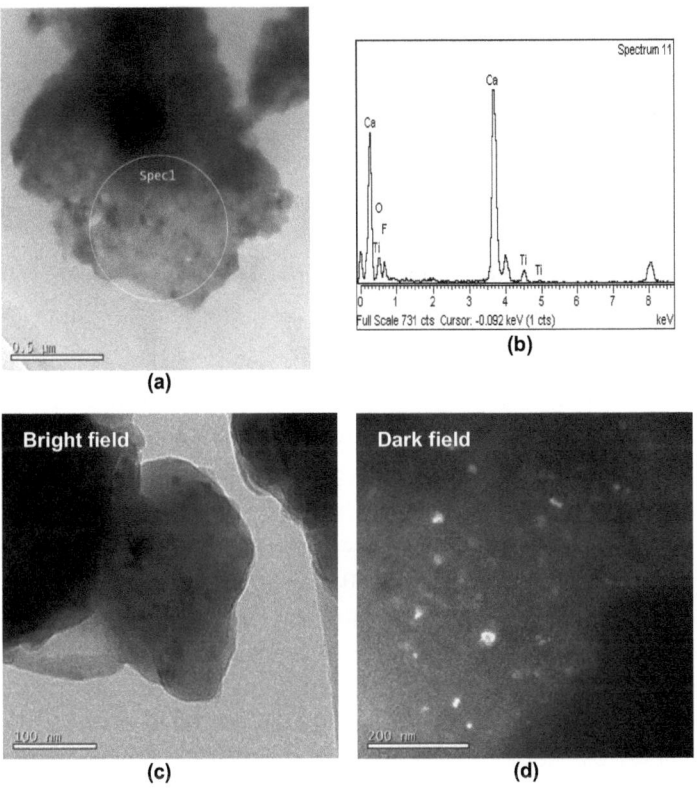

Figure 3.22. TEM images (bright and dark field) together with EDX analysis of the $Ca(BH_4)_2$ sample doped with TiF_4 fully desorbed at 450 °C in vacuum.

Figure 3.22 (a) shows a general picture of the whole sample. The image shows dark spots dispersed within the matrix of the desorbed materials. The EDX analysis (b), performed in the selected area (Spec1) evidences the Ca and Ti signals. The first correspond to the $CaF_{2-x}H_x$, CaO and $CaB_{12}H_{12}$ phase whereas the latter refers either to TiB_2 or Ti_2O_3 nanoparticles. Boron cannot be detected by EDX technique because it represents a too light element. Fluorine and oxygen peaks are visible as well. Selected Area Electron Diffraction could not be performed on the sample due to its instability under the beam. Figures (c) and (d) indicate presence of nanoparticles but EDX cannot clarify whether they are TiB_2 or Ti_2O_3. The oxygen content should be however in the order of contaminations hence not high enough to promote formation of Ti_2O_3 nanoparticles. Formation of TiB_2 nanoparticles would therefore be likely. Picture (d), taken in dark field mode, brings out the nanoparticles highlighting their presence as bright spots. The average nanoparticles size is approximately 20 nm. The lighter grey zone in Figure (d) which embeds the nanoparticles is representative of amorphous phases (e.g. $CaB_{12}H_{12}$).

3.3 Effect of the Ti-isopropoxide on the sorption properties of $Ca(BH_4)_2$

Section 3.2 showed the effectiveness of the TiF_4 and NbF_5 additives in promoting the reversible formation of $Ca(BH_4)_2$. EXAFS and TEM revealed the presence of NbB_2 nanoparticles among the desorption products. Concerning the TiF_4 additive, the identification of the chemical composition of the nanoparticles it is not straightforward. In addition, CaF_2 was formed after milling due to a side reaction between the additive and the borohydride. With the aim to clarify whether the CaF_2 or the transition metal boride nanoparticles are helping the reversible hydrogenation of $Ca(BH_4)_2$, its sorption properties were further investigated by adding either Ti-isopropoxide or CaF_2. The first additive represents a fluorine free compound whereas the latter does not contain any transition metal. In this way, it should be possible to shed some light on the (re)absorption mechanism. Bösenberg et al.[93] as well as Barkhordarian et al.[77] already showed the enhancement of the desorption kinetics when Ti-isopropoxide is employed in combination with RHCs.

3.3.1 The (De)hydrogenation Reaction

The hydrogen desorption analysis of Ca(BH$_4$)$_2$ milled with Ti-isopropoxide was performed heating from room temperature (25 °C) up to 450 °C with a constant heating rate of 3 °C min^{-1} under vacuum (0.02 bar the starting pressure value).

The X-ray diffraction pattern for the milled material is shown in Figure 3.23. The sample indicates simultaneous presence of γ-Ca(BH$_4$)$_2$ and β-Ca(BH$_4$)$_2$ with 7 (± 5 error) and 93 wt. % (± 5 error) abundance respectively. The Figure does not evidence the formation of any Ti-phase or oxide that might have formed by reaction of Ca(BH$_4$)$_2$ with the oxygen groups contained within the organometallic additive.

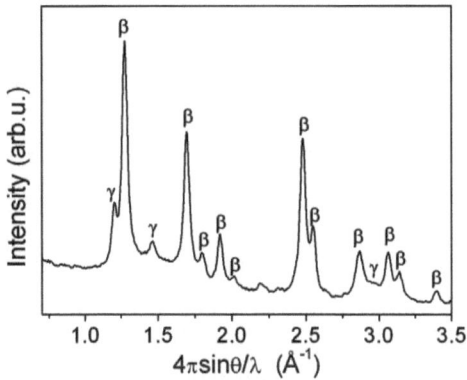

Figure 3.23. SR-PXD pattern at room temperature of Ca(BH$_4$)$_2$ milled with Ti-isopropoxide. γ-Ca(BH$_4$)$_2$ (γ); β-Ca(BH$_4$)$_2$ (β). The measurement was performed at the synchrotron Hasylab, DESY (Hamburg), at the beamline D3.

The first hydrogen desorption reaction of the Ca(BH$_4$)$_2$ + Ti-isopropoxide milled sample is shown in Figure 3.24. The desorption curve for the pure non-milled Ca(BH$_4$)$_2$ is reported as well for comparison. As can be observed in Figure 3.24, the sample milled with Ti-isopropoxide desorbs 7.2 wt. % hydrogen which is less than that released from the pure non-milled Ca(BH$_4$)$_2$ (8.3 wt. %). The lower hydrogen content, in the sample doped with Ti-isopropoxide, might be ascribed to the formation of side products and accompanying hydrogen release during milling due to the addition of the organometallic compound. However, XRD in Figure 3.23 does not evidence the presence of any side product. The main visible effect of the addition of Ti-isopropoxide is the enormous shift to lower desorption temperature compared to the pure non-milled Ca(BH$_4$)$_2$. Figure 3.24 shows that, in the case of the sample milled with Ti-isopropoxide, the first hydrogen desorption step starts already at 210 °C. This desorption reaction proceeds up to 325 °C releasing ca. 1.5 wt. % hydrogen. The final desorption starts at 325 °C and proceeds until the maximum temperature of 450 °C is reached.

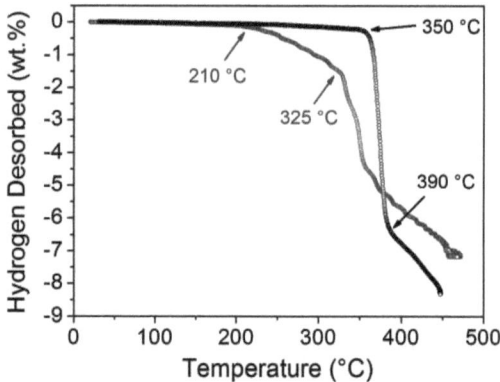

Figure 3.24. Volumetric measurements showing the desorption curves over the temperature of Ca(BH$_4$)$_2$ + Ti-isopropoxide (blue) and pure non-milled Ca(BH$_4$)$_2$ (black). Experiments were carried out by heating the samples from room temperature up to 450 °C in static vacuum (starting value 0.02 bar).

The desorption products were analysed by X-ray diffraction and the pattern is showed in Figure 3.25.

CaH$_2$ and CaO are the only visible phases in Figure 3.25. The Figure does not evidence any trace of CaB$_6$ but, the intense background in the scattering value range of 1.25-2.25 (Å^{-1}) would suggest its existence among the desorption products. The presence of CaO might be responsible for the reduced amount of hydrogen delivered after desorption as shown in Figure 3.24. Note that CaO is often present in the desorption products of Ca(BH$_4$)$_2$ even though the powder handling is performed in protected atmosphere (argon). As mentioned above, Riktor et al.[72] reported the starting Ca(BH$_4$)$_2$ material containing already solvent traces or oxygen-containing impurities.

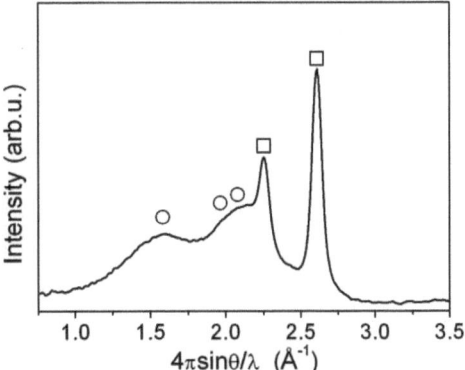

Figure 3.25. SR-PXD pattern of $Ca(BH_4)_2$ milled with Ti-isopropoxide after desorption reaction at 450 °C in vacuum. CaH_2 (O); CaO (□). The measurement was performed at the synchrotron Hasylab, DESY (Hamburg), at the beamline D3.

The desorbed materials were (re)absorbed at 350 °C and 145 bar H_2 for 24 h and subsequently desorbed in the same conditions of the first desorption reaction. The results are reported in section 3.3.3.

3.3.2 Thermal Analysis

With the purpose of investigating the series of events taking place during the decomposition of the $Ca(BH_4)_2$ + Ti-isopropoxide sample, differential scanning calorimetry was carried out. The material was heated from room temperature up to 500 °C with a constant heating rate of 5 °C min^{-1} in 50 ml min^{-1} argon flow. The results are reported in Figure 3.26. For comparison, the DSC curve of the pure non-milled $Ca(BH_4)_2$ (dotted line) is reported as well in the Figure. The onset and peak temperatures for the pure non-milled $Ca(BH_4)_2$ are reported in Figure 3.3 in section 3.1.2.

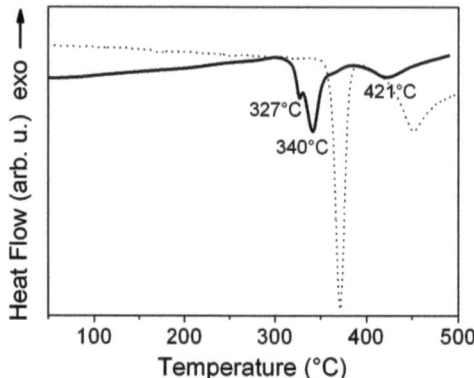

Figure 3.26. DSC curve at 50 ml min^{-1} argon flow of ball milled Ca(BH$_4$)$_2$ + Ti-isopropoxide (solid line); pure non-milled Ca(BH$_4$)$_2$ (dotted line).

The Figure shows three endothermic events whose correspondent peak temperatures are 327, 340 and 421 °C respectively. The first and the second event (at 327 and 340 °C) correspond to the first hydrogen desorption step which, in this case, split in two reactions. The peak at 421 °C corresponds to the second hydrogen desorption step, i.e. the decomposition of Ca$_3$(^{11}BH$_4$)$_3$(^{11}BO$_3$). In the background, as dotted line, the pure non-milled Ca(BH$_4$)$_2$ shows its two step hydrogen desorption reaction. The DSC curve, for the sample milled with Ti-isopropoxide, shows a shift of ca. 30 °C to lower temperatures for both the first and second hydrogen desorption processes compared to the pure non-milled Ca(BH$_4$)$_2$.

3.3.3 The (Re)hydrogenation Reaction

The desorbed materials were (re)absorbed at 350 °C and 145 bar H_2 for 24 h. The XRD pattern after (re)absorption reaction is showed in Figure 3.27.

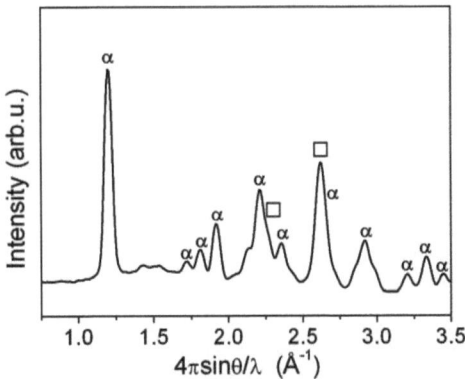

Figure 3.27. SR-PXD pattern of $Ca(BH_4)_2$ milled with Ti-isopropoxide after (re)hydrogenation reaction at 350 °C and 145 bar H_2. α-$Ca(BH_4)_2$ (α); CaO ($\square$). The measurement was performed at the synchrotron Hasylab, DESY (Hamburg), at the beamline D3.

Figure 3.27 evidences the reflections of the α-$Ca(BH_4)_2$ phase showing the successful reversible reaction. The crystallographic peaks of the CaO phase are also present. The oxygen groups contained within the Ti-isopropoxide additive contributed to favour the formation of CaO although oxide phase formation was also observed after decomposition of the pure non-milled $Ca(BH_4)_2$ (Figure 3.4).

The desorption, (re)absorption and subsequent desorption curves for another sample batch of $Ca(BH_4)_2$ with Ti-isopropoxide, obtained by volumetric measurement performed at the Hydrogen Lab at the University of Pavia, are reported in Figure 3.28 A, B and C respectively.

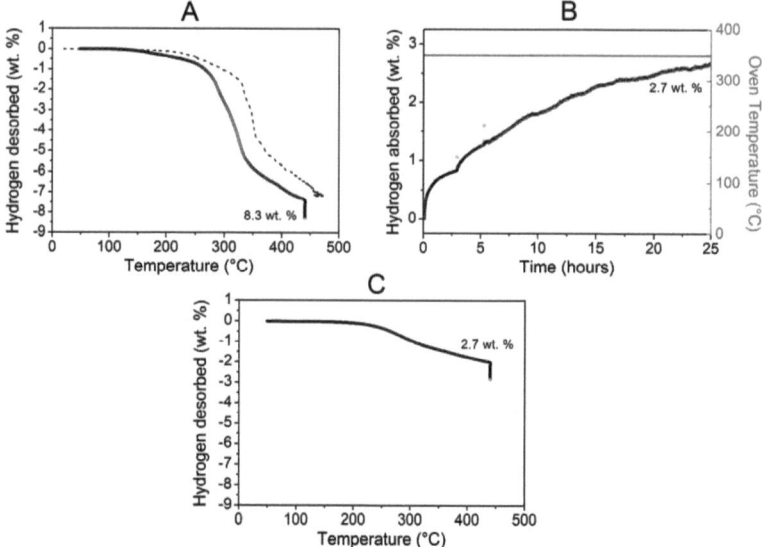

Figure 3.28. Desorption curve of Ca(BH$_4$)$_2$ milled with Ti-isopropoxide performed heating from room temperature up to 450 °C in vacuum (A); subsequent (re)absorption curve of Ca(BH$_4$)$_2$ milled with Ti-isopropoxide performed heating up to 350 °C and 145 bar H$_2$ for 24 hours (B); subsequent desorption curve of Ca(BH$_4$)$_2$ milled with Ti-isopropoxide performed heating from room temperature up to 450 °C in vacuum (C). The measurements were carried out at the Hydrogen Lab at the University of Pavia.

As shown by Figure 3.28 A, the sample desorbs 8 wt. % hydrogen within 3.5 hours which is more than what is shown in Figure 3.24 (7.2 wt. %). The desorption reaction starts at ca. 170 °C releasing ca. 1 wt. % hydrogen. The final desorption starts at 270 °C and proceeds until the maximum temperature of 450 °C is reached.

The (re)absorption curve (Figure 3.28 B) evidences an hydrogen absorption of 2.7 wt. % after 24 hours at 350 °C and 145 H$_2$ pressure. This would correspond to 34 % reversibility. The spikes, observed in Figure B, are caused by the increase of hydrogen pressure (three aliquots of 50 bar each) in order to reach the desired 145 bar H$_2$.

The (re)absorbed sample is subsequently desorbed at 450 °C and vacuum and the resulting curve is reported in Figure 3.28 C. The Figure shows 2.4 wt. % desorption. This value approaches the amount of hydrogen measured for the (re)absorbed sample (Fig. 3.28 C).

The 34 % reversibility, observed for the Ti-isopropoxide doped Ca(BH$_4$)$_2$ sample, is close to the value measured for the TiF$_4$ doped sample (35 %). Instead, NbF$_5$ doped sample still exhibits better performance (55 % reversibility).

The products after second hydrogen desorption were analysed by X-ray diffraction and the pattern is showed in Figure 3.29.

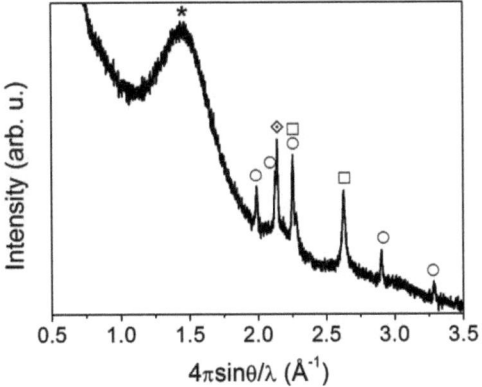

Figure 3.29. XRD pattern of Ca(BH$_4$)$_2$ milled with Ti-isopropoxide after the second desorption reaction (Figure 3.28 C) at 450 °C in vacuum. CaH$_2$ (O); CaO (□); CaB$_6$ (◇); plastic cap of the sample holder (*). The measurement was performed at the Hydrogen Lab at the University of Pavia.

The diffractogram evidences presence of CaH$_2$, CaO and CaB$_6$. After first desorption reaction (Fig. 3.25) CaB$_6$ is not visible. As can be observed in Figure 3.29, the most intense signal of CaB$_6$ falls between the CaH$_2$ peaks. The XRDs after first desorption (Fig. 3.25), shows broad peaks of CaH$_2$ which probably merge the hexaboride most intense signal.

3.3.4 $^{11}B\{^1H\}$ Magic Angle Spinning–Nuclear Magnetic Resonance

XRD of the material after first hydrogen desorption reaction (Figure 3.25) shows presence of CaH_2 and CaO. However, the diffractogram does not evidence presence of CaB_6. Its existence is clearly indicated after the second hydrogen desorption reaction (Figure 3.29).

$^{11}B\{^1H\}$ MAS-NMR was carried out on the powder after first hydrogen desorption reaction with the aim to identify the nature of the final B-containing compounds. The spectrum of the desorbed material is shown in Figure 3.30 A together with those of the reference compounds. $^{11}B\{^1H\}$ NMR was performed at room temperature.

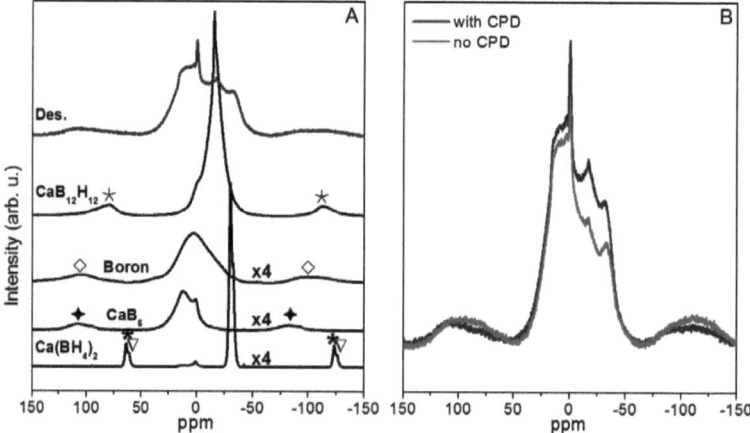

Figure 3.30. A: $^{11}B\{^1H\}$ MAS-NMR spectra at room temperature of $Ca(BH_4)_2$ milled with Ti-isopropoxide after first hydrogen desorption reaction at 450 °C in vacuum. $Ca(BH_4)_2$ purchased from Sigma-Aldrich, CaB_6 and boron, scale adjusted by ¼. Side bands are indicated by ∗, ∇, ◆, ◊, ⋆. B: $^{11}B\{^1H\}$ MAS-NMR spectra at room temperature of $Ca(BH_4)_2$ milled with Ti-isopropoxide after first hydrogen desorption reaction at 450 °C in vacuum collected with and without Composite-Pulse Decoupling (CPD).

The spectrum of desorbed $Ca(BH_4)_2$ + Ti isopropoxide material (blue spectrum) shows a series of four signals: ~+12 ppm, ~-0.6 ppm, ~-17 ppm and ~-32 ppm. By comparison with the spectra of the reference compounds we can associate the peak at +12 ppm to CaB_6 and the one at -30 ppm to residual β-$Ca(BH_4)_2$. The peak at ~-17 ppm corresponds to $CaB_{12}H_{12}$. The signals of the desorbed material are slightly shifted respect to the values observed for the reference compounds. The values might shift in dependence of the disordered nature of the phase structure.[90] The narrow signal at -0.6 ppm falls in the range of chemical shift corresponding to a BH_3-L (L=ligand) compound.[116] The profile of the NMR spectrum in Figure 3.30 A for the desorbed material, apart from the peak at -0.6 ppm, is very similar to those of the desorbed NbF_5 and TiF_4 doped $Ca(BH_4)_2$ material. This signal was never observed in the case of the transition metal fluoride doped samples. It is worth to say that the $^{11}B\{^1H\}$ MAS-NMR analysis was performed on a one-year old sample and although material storage was performed in the best conditions possible, contamination by moisture cannot be completely excluded in one-year time. In addition, it was already reported that oxygen contamination can be already existing in the starting $Ca(BH_4)_2$ powder.[72] Nevertheless, to clarify the nature of the bond characterising the compound giving a signal at -0.6 ppm, the NMR experiment was repeated, on the same desorbed powder, without Composite-Pulse Decoupling (CPD). The result is reported in Figure 3.30 B. If the material contains a B-H bond a considerable decrease of the intensity of its peak should be observed. By contrast, if it does not contain B-H bond, the intensity of the peak should slightly decrease. Figure 3.30 B highlights how the peaks at ~-17 ppm, at ~-32 ppm and at -0.6 ppm decrease their intensity by a similar proportional value whereas the signal at ~+12 ppm does not show the same decrement. The latter signal (+12 ppm) belongs to the CaB_6 phase and since it does not contain B-H bonds its intensity, in both experiments, is comparable. The peaks at ~-17 ppm and at ~-32 ppm are known containing B-H bonds ($CaB_{12}H_{12}$ and β-$Ca(BH_4)_2$ respectively). The NMR experiment confirms that the signal at -0.6 ppm belongs to a BH_3-L (L=ligand) compound. Unfortunately, the identification of the ligand is not easy and, considering the fact that it might be due to contamination, it is not relevant.

3.3.5 Transmission Electron Microscopy

Since no EXAFS experiment was performed on $Ca(BH_4)_2$ samples doped with Ti-isopropoxide, TEM microscopy was carried out in order to confirm the presence of the Ti_2O_3 or transition metal boride nanoparticles (TiB_2).

The TEM pictures for the (re)absorbed Ti-isopropoxide doped sample are reported in Figure 3.31 together with EDX and SAED analysis. The detection of the nanoparticles was complex due to the instability of the material under the electron beam. The powder decomposes within a few seconds.

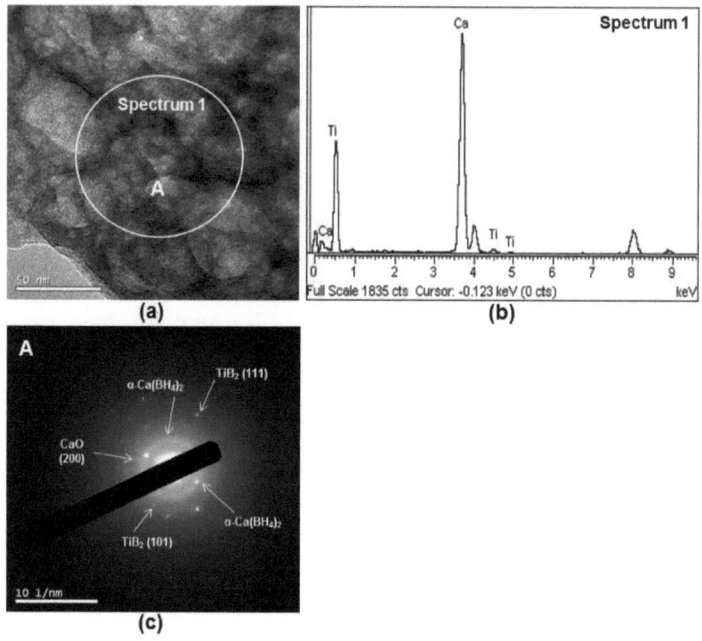

Figure 3.31. TEM images (bright field) together with EDX analysis and SAED of the Ca(BH$_4$)$_2$ sample doped with Ti-isopropoxide (re)absorbed at 350 °C and 145 bar H$_2$.

Figure 3.31 (a) shows a general overview of the sample. The EDX analysis (Fig. 3.31 b), performed in the selected area (Spectrum1) evidences the Ca and Ti signals. No titanium phase was determined by XRD in the (re)absorbed material (Fig. 3.27). Ca signals can be attributed to α-Ca(BH$_4$)$_2$ and CaO phase (Fig. 3.27). SAED, in Figure 3.31 (c), performed in the area indicated as A (Figure 3.31 (a)) shows the spots of the α-Ca(BH$_4$)$_2$, CaO and TiB$_2$ phase. Unfortunately, without dark field mode image, it is not possible to observe the transition metal boride nanoparticles and thus determine the particle size. However, since XRD (Fig. 3.27), also in this case, does not show the reflections of TiB$_2$, one can reasonably assume the nanoparticles being around 10 nm.

3.4 Effect of the CaF$_2$ additive on the sorption properties of Ca(BH$_4$)$_2$

Section 3.2 and 3.3 showed the positive influence of the transition metal fluoride additives (TiF$_4$ and NbF$_5$) as well as the Ti-isopropoxide in promoting the reversible formation of Ca(BH$_4$)$_2$. In order to define the (re)absorption reaction mechanism, the last set of experiments involve the investigation of the effect of CaF$_2$ on the reversible formation of Ca(BH$_4$)$_2$.

3.4.1 The (De)hydrogenation Reaction

The hydrogen desorption analysis of Ca(BH$_4$)$_2$ milled with CaF$_2$ was performed heating from room temperature (25 °C) up to 450 °C with a constant heating rate of 3 °C min^{-1} under vacuum (0.02 bar the starting pressure value).

The X-ray diffraction pattern for the milled material is presented in Figure 3.32. Rietveld analysis on the sample after milling indicates presence of γ-Ca(BH$_4$)$_2$, β-Ca(BH$_4$)$_2$ and CaF$_2$ with 58 (± 5 error), 35 (± 5 error) and 7 wt. % (± 5 error) phase abundance respectively.

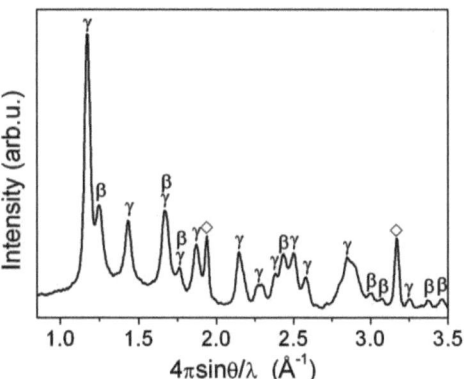

Figure 3.32. SR-PXD pattern of Ca(BH$_4$)$_2$ milled with CaF$_2$. γ-Ca(BH$_4$)$_2$ (γ); β-Ca(BH$_4$)$_2$ (β); CaF$_2$ (◇). The measurement was performed at the synchrotron Hasylab, DESY (Hamburg), at the beamline D3.

The first hydrogen desorption reaction, for the Ca(BH$_4$)$_2$ + CaF$_2$ milled sample, is shown in Figure 3.33. The desorption curve for the pure non-milled Ca(BH$_4$)$_2$ is reported as well for comparison purposes.

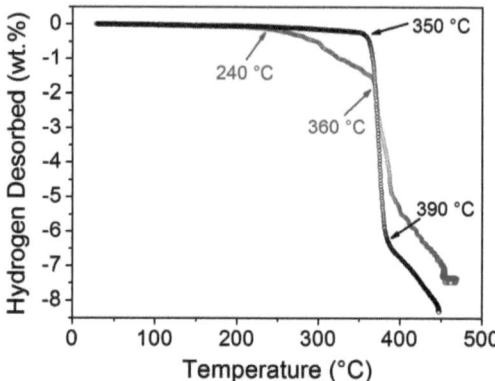

Figure 3.33. Volumetric measurements showing the desorption curves over the temperature of Ca(BH$_4$)$_2$ + CaF$_2$ (green) and pure non-milled Ca(BH$_4$)$_2$ (black). Experiments were carried out by heating the samples from room temperature up to 450 °C in static vacuum (starting value 0.02 bar).

As can be observed in Figure 3.33, the sample milled with CaF$_2$ desorbs 7.2 wt. % hydrogen which is lower compared to the pure non-milled Ca(BH$_4$)$_2$ (8.3 wt. %). The lower amount of hydrogen released is linked to the reduced quantity of Ca(BH$_4$)$_2$ available because of the addition of CaF$_2$. CaF$_2$ contributes to shift to lower temperature values the desorption processes respect to the pure non-milled Ca(BH$_4$)$_2$. Figure 3.33 shows that, for the sample milled with CaF$_2$, the first hydrogen desorption step starts already at 240 °C. This desorption reaction proceeds slowly up to 360 °C releasing ca. 1.8 wt. % hydrogen. In the 240-360 °C temperature range, the desorption reaction does not proceed smoothly as can be observed by the variation of the slope in Figure 3.33. The final desorption step begins at 360 °C and proceeds until plateau.

The products of the desorption reaction were analysed by X-ray diffraction and the pattern is showed in Figure 3.34.

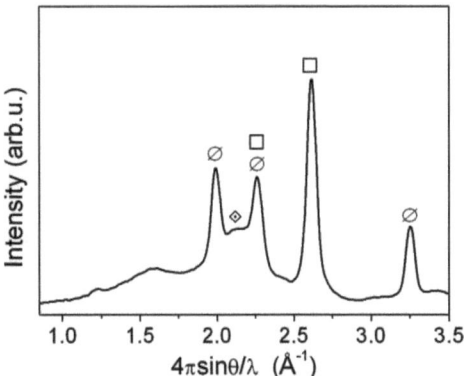

Figure 3.34. SR-PXD pattern of Ca(BH$_4$)$_2$ milled with CaF$_2$ after desorption reaction at 450 °C in vacuum. CaF$_{2-x}$H$_x$ (∅); CaO (□); CaB$_6$ (◇). The measurement was performed at the synchrotron Hasylab, DESY (Hamburg), at the beamline D3.

Figure 3.34 evidences the peaks of the CaF$_{2-x}$H$_x$, CaO and CaB$_6$ phase. CaF$_{2-x}$H$_x$ was showed already to be a product of the desorption reaction between CaH$_2$ and CaF$_2$ (reactions 1 and 2 section 3.2.3). Its presence was observed after decomposition in vacuum for the samples milled with TiF$_4$ and NbF$_5$. Figure 3.34 reports again the presence of CaO phase although the powder handling was performed in argon atmosphere but Riktor et al.[72] clarified the reasons of the contamination. Furthermore, the Figure shows a broad amorphous like peak around 1.5 Q (Å^{-1}) which suggests the presence of nano-crystalline or amorphous phases. Rietveld analysis of the pattern showed in Figure 3.34 was not performed. The presence of amorphous phases which cannot be detected by XRD, would contribute to overestimate the values of the phase abundance.

The desorbed materials were (re)absorbed at 350 °C and 145 bar H$_2$ for 24 h and the result will be reported in the section 3.4.2.

3.4.2 Thermal Analysis

With the purpose of investigating the series of events taking place during the decomposition of Ca(BH$_4$)$_2$ + CaF$_2$ sample, a differential scanning calorimetry measurement was carried out. The material was investigated by heating up from room temperature up to 500 °C with a constant heating rate of 5 °C min^{-1} in 50 ml min^{-1} argon flow. The results are reported in

Figure 3.35. For comparison purposes the DSC curve of the pure non-milled Ca(BH$_4$)$_2$ (dotted line) is reported as well in the Figure. The onset and peak temperatures for the pure non-milled Ca(BH$_4$)$_2$ are reported in Figure 3.3 in the section 3.1.2.

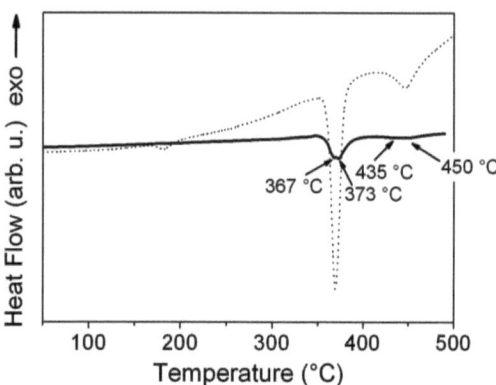

Figure 3.35. DSC curve at 50 ml min^{-1} argon flow of ball milled Ca(BH$_4$)$_2$ + CaF$_2$ (solid line); pure non-milled Ca(BH$_4$)$_2$ (dotted line).

The Figure shows two endothermic events with peak temperatures at 367 and 373 °C respectively. Other two endothermic reactions take place around 435 and 450 °C, respectively. Both series of events are related to the first and second hydrogen desorption step respectively. The first desorption step splits in two consecutive reactions as well as the second step. In Figure 3.35, the second desorption reaction becomes broader, compared to that of the non-milled material (dotted curve). This effect was already observed in the DSC curves in Figure 3.12 B for the Ca(BH$_4$)$_2$ milled with transition metal fluorides.

3.4.3 The (Re)hydrogenation Reaction

The desorbed materials (CaF$_{2-x}$H$_x$, CaB$_6$ and CaO) were (re)absorbed at 350 °C and 145 bar H$_2$ for 24 h. The XRD pattern after (re)absorption reaction is showed in Figure 3.36. The absorption curve, obtained by volumetric measurement, will not be reported here due to its low quality which makes the interpretation of the data uncertain.

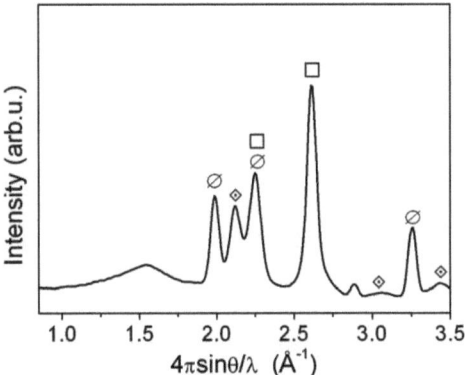

Figure 3.36. SR-PXD pattern of Ca(BH$_4$)$_2$ milled with CaF$_2$ after (re)absorption reaction at 350 °C and 145 bar H$_2$ pressure. CaF$_{2-x}$H$_x$ (∅); CaO (□); CaB$_6$ (◇). The measurement was performed at the synchrotron Hasylab, DESY (Hamburg), at the beamline D3.

Figure 3.36 still shows the crystallographic peaks of the (de)hydrogenated phases (CaF$_{2-x}$H$_x$, CaO and CaB$_6$) evidencing that no (re)absorption reaction has occurred. In Figure 3.36, the peak at 2.12 Q (Å^{-1}) of the CaB$_6$ phase is more intense compared to Figure 3.34. The long exposure time (24h) to both high temperature and pressure influences the peaks coarsening. Furthermore, the broad peak around 1.5 Q (Å^{-1}) is still present and it might suggest the existence of nano-crystalline or amorphous-like phases.

3.5 Ca(BH$_4$)$_2$ + MgH$_2$

The hydrogen desorption process of the Ca(BH$_4$)$_2$ + MgH$_2$ composite system was characterised by means of volumetric measurements, calorimetric techniques, *in* and *ex-situ* XRD analysis, [11]B Magic Angle Spinning-Nuclear Magnetic Resonance and Transmission Electron Microscopy.

3.5.1 The (De)hydrogenation Reaction

The volumetric analysis representing the first hydrogen desorption reaction of the ball milled Ca(BH$_4$)$_2$ + MgH$_2$ system was performed in a Sievert's type apparatus heating from room temperature (25 °C) up to 400 °C with a constant heating rate of 3 °C min^{-1} under vacuum (0.02 bar starting pressure).

The XRD pattern of the starting Ca(BH$_4$)$_2$ + MgH$_2$ milled material is reported in Figure 3.37. The Figure indicates the presence of the low temperature γ (space group Pbca, orthorhombic phase)[102], of the high temperature β-Ca(BH$_4$)$_2$ (space group P4, tetragonal phase)[71] and of the MgH$_2$ phase. The narrow peaks in Figure 3.36 describe the crystalline status of all the phases.

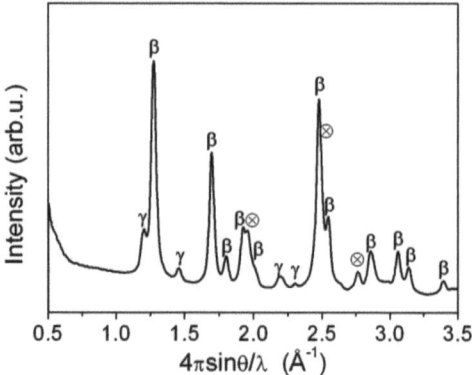

Figure 3.37. SR-PXD pattern at room temperature of milled Ca(BH$_4$)$_2$ + MgH$_2$ composite system. γ-Ca(BH$_4$)$_2$ (γ); β-Ca(BH$_4$)$_2$ (β); MgH$_2$ (⊗). The measurement was performed at the synchrotron MAX-lab, Lund (Sweden) at the beamline I711.

The relative phase abundance, calculated by Rietveld method (by MAUD software)[87], is 10 wt. % (± 5 error), 73 wt. % (± 5 error) and 17 wt. % (± 5 error) for γ-Ca(BH$_4$)$_2$, β-Ca(BH$_4$)$_2$ and MgH$_2$ respectively.

The volumetric analysis of the first hydrogen desorption reaction is reported in Figure 3.38. The kinetic curve evidences two different slopes referred to two distinct desorption processes. The first hydrogen desorption reaction starts at ca. 350 °C whereas the second begins at ca. 375 °C.

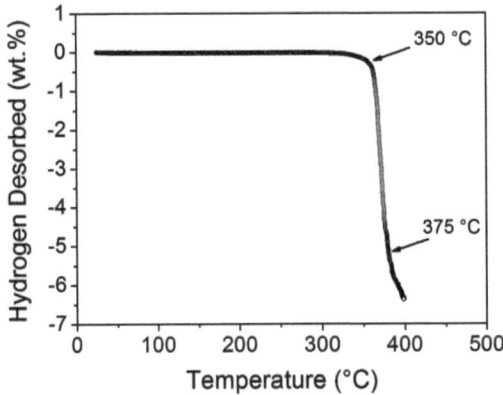

Figure 3.38. Volumetric measurements showing the desorption curve of the Ca(BH$_4$)$_2$ + MgH$_2$ composite (black) over the temperature. The experiment was carried out by heating the sample from room temperature up to 400 °C in static vacuum (starting value 0.02 bar).

Figure 3.38 shows that 6.4 wt. % hydrogen is released within 30 min in the 350–400 °C temperature range. In theory, Ca(BH$_4$)$_2$ + MgH$_2$ composite system contains 10.5 wt. %. However, reactions 12, 13 and 14 in section 1.2 indicate that the amount of hydrogen desorbed varies in dependence of the decomposition path. For these reactions, the theoretical amount of hydrogen desorbed is 8.4, 9.1 and 8.4 wt. % respectively. Besides CaH$_2$, among the products for all the aforementioned processes, reaction 12 leads to the formation of MgB$_2$. Its presence after (de)hydrogenation could be an indication for the simultaneous (de)composition of Ca(BH$_4$)$_2$ and MgH$_2$. The reaction involving the formation of MgB$_2$ is representative of the Reactive Hydride Composites concept. In case of reactions 13 and 14, Ca(BH$_4$)$_2$ and MgH$_2$

react separately and lead, besides CaH_2, to the formation of CaB_6 or B and Mg respectively. The amount of hydrogen reported for the first hydrogen desorption in Figure 3.38 (6.4 wt. %) is much lower than the theoretical value corresponding to the reactions 12, 13 and 14. Formation of side products upon decomposition might be responsible for the lower capacity observed. *In-situ* X-ray diffraction together with ^{11}B MAS-NMR will shed some light on this point.

3.5.2 Thermal Analysis

With the purpose of investigating the ongoing series of events taking place during the decomposition of ball milled $Ca(BH_4)_2$ + MgH_2 composites, differential scanning calorimetry and thermogravimetry were carried out. The sample was measured while heating from room temperature up to 500 °C with a constant heating rate of 5 °C min^{-1} in 50 ml min^{-1} argon flow. The results are reported in Figure 3.39.

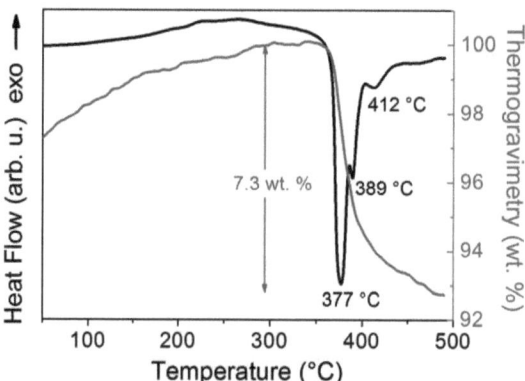

Figure 3.39. DSC (black) and thermogravimetric (red) curve of ball milled $Ca(BH_4)_2$ + MgH_2 composite at 50 ml min^{-1} argon flow with an heating rate of 5 °C min^{-1}.

Figure 3.39 shows three different endothermic events with peak temperatures of 377, 389 and 412 °C respectively. The thermogravimetry shows a 7.3 wt. % weight loss for the whole decomposition reaction. The first endothermic event shows an onset temperature of ca. 350 °C matching that one observed for the volumetric measurements in Figure 3.38. The three series of events showed by the DSC curve, might represent a non-simultaneous decomposition of the Ca(BH$_4$)$_2$ and MgH$_2$ in the composite.

The nature of the reactions occurring during decomposition of the Ca(BH$_4$)$_2$ + MgH$_2$ milled composite system was studied by means of *in-situ* XRD.

3.5.3 *In-situ* Synchrotron Radiation Powder X-ray Diffraction

In-situ Synchrotron Radiation Powder X-ray diffraction was carried out on the Ca(BH$_4$)$_2$ + MgH$_2$ composite in order to understand the reaction mechanism upon decomposition reaction. The diffraction patterns are reported over the temperature in Figure 3.40. The powder, contained in a sapphire capillary, was heated from room temperature (25 °C) up to 400 °C in static vacuum with a constant heating rate of 5 °C min^{-1}.

The diffraction data at 30 °C contains the γ-, β-Ca(BH$_4$)$_2$ and the MgH$_2$ phase. Their relative abundance is already reported in section 3.5.1. In the temperature range of 30-271 °C the low temperature γ-Ca(BH$_4$)$_2$ gradually transforms into the high temperature β-Ca(BH$_4$)$_2$ phase. At 271 °C, β-Ca(BH$_4$)$_2$ and MgH$_2$ are the only phases present. They coexist without any further change in the 271-340 °C temperature range. At 355 °C the reflections of the Mg and Ca$_4$Mg$_3$H$_{14}$ [117] phase are visible. At this temperature, the peaks of the β-Ca(BH$_4$)$_2$ and MgH$_2$ phase are still present evidencing a lower intensity compared to the pattern at 340 °C. The ternary Ca-Mg-H phase decomposes further, in the 373–400 °C temperature range, to CaH$_2$, Mg and H$_2$ (reaction 1):

(1) Ca$_4$Mg$_3$H$_{14}$ ↔ 4 CaH$_2$ + 3 Mg + 3 H$_2$

As shown by the DSC in Figure 3.39, 355 °C (377 °C peak temperature) and 389 °C correspond to the onset and to the peak temperature of the first and of the second endothermic event respectively.

The *in-situ* XRD analysis in Figure 3.40 cannot clearly show which phase between β-Ca(BH$_4$)$_2$ and MgH$_2$ desorbs as first. Their decomposition temperature is very close (within 20 °C temperature range). In order to shed some light on this regard, a detailed comparison of the XRD patterns belonging to the *in-situ* experiment reported in Figure 3.40 is provided in Figure 3.41, 3.42 and 3.43 for several temperatures.

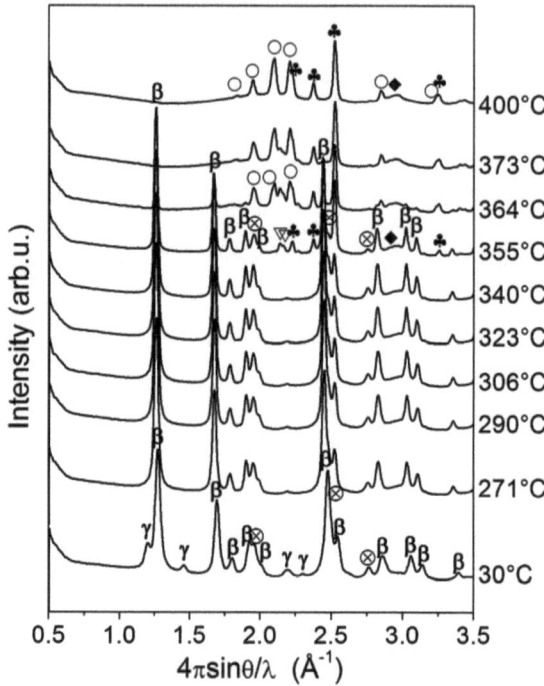

Figure 3.40. SR-PXD patterns of ball milled Ca(BH$_4$)$_2$ + MgH$_2$ composite. The experiment was carried out by heating in vacuum from RT up to 400 °C with 5 °C min^{-1} constant heating rate. γ-Ca(BH$_4$)$_2$ (γ); β-Ca(BH$_4$)$_2$ (β); MgH$_2$ (⊗); Ca$_4$Mg$_3$H$_{14}$ (∇); Mg (♣); CaH$_2$ (○); MgO (◆). The measurement was performed at the synchrotron MAX-lab, Lund (Sweden) at the beamline I711.

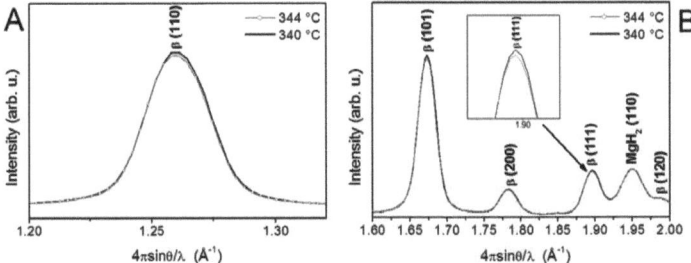

Figure 3.41. SR-PXD patterns at 340 and 344 °C of ball milled Ca(BH$_4$)$_2$ + MgH$_2$ composite. The experiment was carried out by heating the sample in vacuum from RT up to 400 °C with 5 °C min^{-1} as constant heating rate. β-Ca(BH$_4$)$_2$ (β). Phase lattice planes are reported in brackets. The measurement was performed at the synchrotron MAX-lab, Lund (Sweden) at the beamline I711. (A): scattering vector value range of 1.20-1.32 (Å^{-1}); (B): scattering vector value range of 1.60-2.00 (Å^{-1}).

Figure 3.41 A and B report the XRD patterns of the Ca(BH$_4$)$_2$ + MgH$_2$ composite at 340 and 344 °C for two scattering vector value ranges: 1.20-1.32 (Å^{-1}) and 1.60-2.00 (Å^{-1}). The scattering vector range is divided in two parts in order to better evaluate the small changes of peak intensities.

Figure 3.41 A and B shows the decrease of intensity of the (110), (101) and (111) lattice plane for the β-Ca(BH$_4$)$_2$ phase in the 340-344 °C temperature range. Instead, no variation of scattering intensity is detected for the MgH$_2$ phase (110) in the same temperature range. Its XRD patterns at 340 and 344 °C overlap. This result confirms that, in our study, Ca(BH$_4$)$_2$ starts decomposing before MgH$_2$.

Figure 3.42 A, B and C report the XRD patterns of the Ca(BH$_4$)$_2$ + MgH$_2$ composite at 352, 356 and 360 °C for three scattering vector value ranges respectively: 1.20-1.35 (Å^{-1}), 1.60-2.00 (Å^{-1}) and 2.00-2.50 (Å^{-1}).

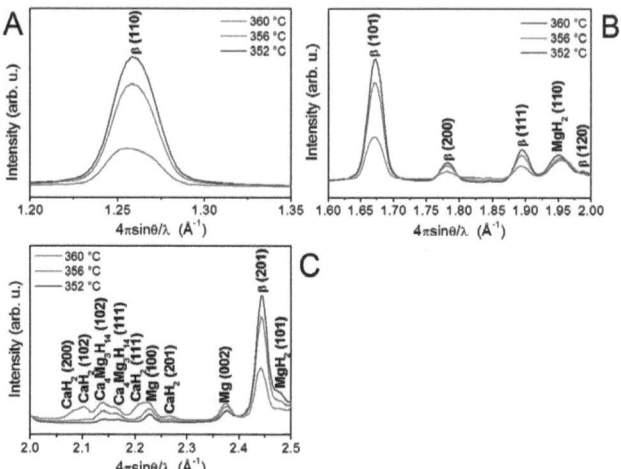

Figure 3.42. SR-PXD patterns at 352, 356 and 360 °C of ball milled Ca(BH$_4$)$_2$ + MgH$_2$ composite. The experiment was carried out by heating in vacuum from RT up to 400 °C with 5 °C min^{-1} as constant heating rate. β-Ca(BH$_4$)$_2$ (β). Phase lattice planes are reported in brackets. The measurement was performed at the synchrotron MAX-lab, Lund (Sweden) at the beamline I711. (A): scattering vector value range of 1.20-1.35 (Å^{-1}); (B): scattering vector value range of 1.60-2.00 (Å^{-1}); (C): scattering vector value range of 2.00-2.50 (Å^{-1}).

The XRD patterns in Figure 3.42 A and B show that the peak intensities of β-Ca(BH$_4$)$_2$ and MgH$_2$ decrease when the temperature increases from 352 up to 360 °C. The reflections of β-Ca(BH$_4$)$_2$ reduce their intensity faster compared to MgH$_2$. The pattern at 352 °C in Figure 3.42 C evidences the Bragg peaks of the Mg phase ((100) and (002)). In the same scattering vector range, the (102) and (111) lattice planes of the Ca$_4$Mg$_3$H$_{14}$ phase appear. Kim et al.[48] reported that, during desorption, the formation of Ca$_4$Mg$_3$H$_{14}$ can be promoted either by reaction between Mg and Ca(BH$_4$)$_2$ or by CaH$_2$ (obtained by decomposition of Ca(BH$_4$)$_2$) Mg and H$_2$. During hydrogenation, Kim et al.[118] observed that CaH$_2$ reacts together with MgH$_2$ to form Ca$_4$Mg$_3$H$_{14}$ as shown by reaction 2.

(2) $\quad$ 4 CaH$_2$ + 3 MgH$_2$ → Ca$_4$Mg$_3$H$_{14}$

Figure 3.42 C evidences how the $Ca_4Mg_3H_{14}$ phase fraction grows in the 352-360 °C temperature range. At 360 °C, CaH_2 is formed as can be observed by the appearance of several of its lattice planes in the 2.00-2.50 ($Å^{-1}$) scattering vector range.

Figure 3.43 A, B and C report the XRD patterns of the $Ca(BH_4)_2$ + MgH_2 composite at 364 and 368 °C for three scattering vector value ranges respectively: 1.20-1.35 ($Å^{-1}$), 1.60-2.00 ($Å^{-1}$) and 2.00-2.50 ($Å^{-1}$).

Figures 3.43 A, B and C show the absence of MgH_2 reflections which is completely desorbed at these temperatures. The diffractograms in Figure 3.43 A and B show that, at 364 °C residual β-$Ca(BH_4)_2$ is still present. At 368 °C, its phase fraction goes to zero evidencing the end of its decomposition reaction. Therefore, although β-$Ca(BH_4)_2$ starts its desorption before MgH_2, the MgH_2 phase desorbs hydrogen much faster than β-$Ca(BH_4)_2$. In this temperature range, CaH_2 (011), (102) and (111) lattice planes increase their intensities. The increase of the CaH_2 phase fraction originates from the residual β-$Ca(BH_4)_2$. In Figure 3.43 C, for the 364-368 °C temperature range, the $Ca_4Mg_3H_{14}$ and the Mg phase fractions are constant. This is due to the absence of MgH_2 (fully desorbed at this temperature) which cannot react anymore with the CaH_2 delivered by the residual β-$Ca(BH_4)_2$ in order to produce $Ca_4Mg_3H_{14}$. This result explains why the Mg and $Ca_4Mg_3H_{14}$ scattering intensities remain constant. These observations indicate that the formation of Ca-Mg-H follows reaction 2.

First principle calculations showed that $Ca_4Mg_3H_{14}$ decomposes at higher temperatures than MgH_2 because the Gibbs energy change for reaction 1 is slightly negative (-7.9 kJ at 350 °C).[48] The final products of the whole desorption reaction are CaH_2, Mg and MgO. MgB_2 is not visible in the XRD pattern at 400 °C as well as other boron-phases (e.g. boron, CaB_6, $CaB_{12}H_{12}$) implying their amorphous or nanocrystalline status as already observed in the case of the transition metal fluorides doped $Ca(BH_4)_2$ system. To detect those phases, $^{11}B\{^1H\}$ MAS-NMR experiments was performed on the (de)hydrogenated samples.

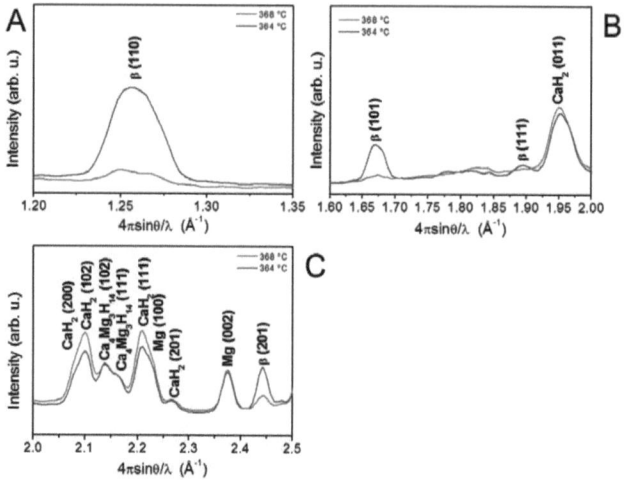

Figure 3.43. SR-PXD patterns at 364 and 368 °C of ball milled Ca(BH$_4$)$_2$ + MgH$_2$ composite. The experiment was carried out by heating the sample in vacuum from RT up to 400 °C with 5 °C min^{-1} as constant heating rate. β-Ca(BH$_4$)$_2$ (β). Phase lattice planes are reported in brackets. The measurement was performed at the synchrotron MAX-lab, Lund (Sweden) at the beamline I711. (A): scattering vector value range of 1.20-1.35 (Å^{-1}); (B): scattering vector value range of 1.60-2.00 (Å^{-1}); (C): scattering vector value range of 2.00-2.50 (Å^{-1}).

3.5.4 The (Re)hydrogenation Reaction

The products of desorption shown in section 3.5.3 (CaH$_2$, Mg, MgO and boron-phase) were (re)absorbed at 350 °C and 145 bar H$_2$ for 24 and 43 hours respectively. Such a long isothermal time was applied considering the generally known slow absorption reaction kinetics of tetrahydroborates. The absorption curve will not be reported here due to its low quality that does not provide any clear information about the quantity of hydrogen (re)absorbed. X-ray diffraction was performed on the (re)absorbed material in order to determine whether the (re)absorption reaction was successful. A subsequent desorption volumetric measurement, on the (re)absorbed material, was carried out to determine the amount of hydrogen reversibly absorbed. Figure 3.44 shows the XRD patterns for the samples (re)hydrogenated for 24 and 43 hours.

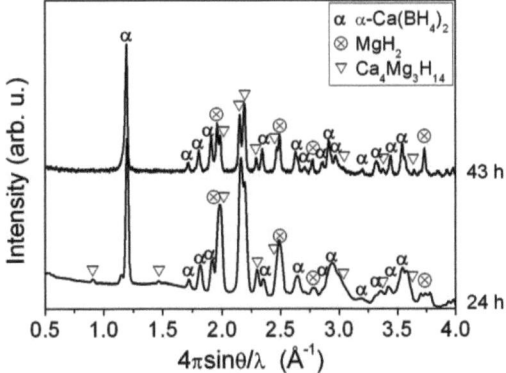

Figure 3.44. XRD spectra of the (re)hydrogenated powders after absorption for 24 and 43 hours at 350 °C and 145 bar H_2. α-$Ca(BH_4)_2$ (α); MgH_2 ($\otimes$); $Ca_4Mg_3H_{14}$ (∇). The measurements were performed at the Institute for Metallic Materials at the Leibniz Institute for Solid State and Materials Research (Dresden).

The Figure shows, besides MgH_2, the reflections of the α-$Ca(BH_4)_2$ phase evidencing the successful (re)hydrogenation reaction. However, presence of $Ca_4Mg_3H_{14}$ phase, in both the XRD patterns after 24 and 43 hours, demonstrates that only partial reversibility was achieved if the formation of $Ca(BH_4)_2$ is considered. The crystallographic peaks of the $Ca_4Mg_3H_{14}$ phase, in the pattern after 43 hours, show a lower intensity compared to those in the pattern after 24 hours. This demonstrates that the $Ca_4Mg_3H_{14}$ phase is slowly consumed during (re)absorption reaction.

$^{11}B\{^1H\}$ MAS-NMR measurements (shown in 3.5.5) were carried out on desorbed and (re)absorbed materials in order to get information on the status of boron and on the (re)absorption reaction mechanism in the $Ca(BH_4)_2$ + MgH_2 system.

The curves for the second and third hydrogen desorption processes are reported in Figure 3.45.

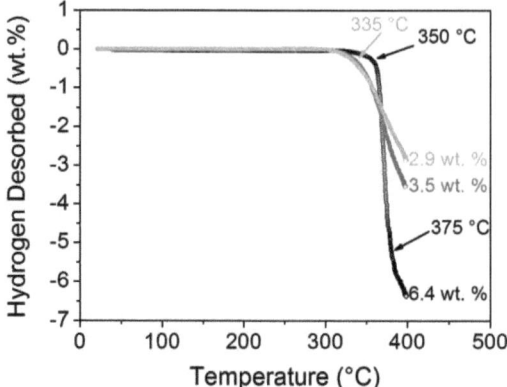

Figure 3.45. Volumetric measurements showing the 1st (black), 2nd (gray) and 3rd (light gray) hydrogen desorption curves over the temperature for the Ca(BH$_4$)$_2$ + MgH$_2$ composite. The experiments were carried out by heating the sample from room temperature up to 400 °C in static vacuum (starting value 0.02 bar). (Re)absorption reactions were performed at 350 °C and 145 bar H$_2$ for 24 hours.

The Figure shows that, after (re)hydrogenation for 24 hours at 350 °C and 145 bar H$_2$, 3.5 wt. % hydrogen is desorbed under the applied conditions. Taking the quantity of hydrogen evolved during the first desorption reaction (6.4 wt. %) as basis, 55 % of the reaction is going on in a reversible manner. The amount of hydrogen delivered during the third desorption is decreased again to 2.9 wt. % only. In this case, the ratio of reversibility is 83 % if compared to the 2nd desorption. Therefore, comparing the values for the reversibility observed during the first, second and third hydrogen desorption, the formation of potential stable compounds (e.g. amorphous boron, CaB$_{12}$H$_{12}$) which limits the reversibility and does not contribute anymore in the following hydrogenation cycles mainly occurs during the first desorption process.

3.5.5 ^{11}B{^{1}H} Magic Angle Spinning–Nuclear Magnetic Resonance

Since no boron-phase was detected by XRD after desorption, both the (de)hydrogenated and (re)hydrogenated samples were analysed by ^{11}B{^{1}H} MAS-NMR. The NMR spectrum of the desorbed material is shown in Figure 3.46 together with those of selected reference compounds.

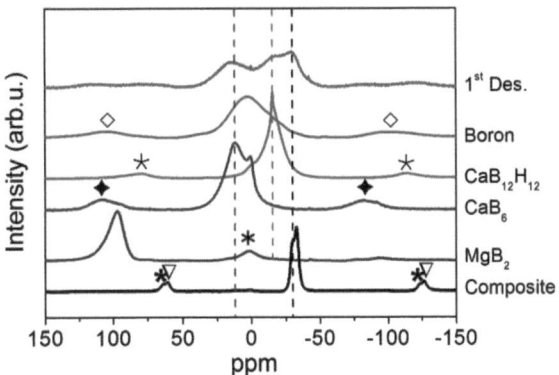

Figure 3.46. ^{11}B{^{1}H} MAS-NMR spectrum at room temperature of milled Ca(BH$_4$)$_2$ + MgH$_2$ composite desorbed at 400 °C in vacuum (1st Des.). Side bands are indicated by ★, ▽, ∗, ♦, ⋆, ◊.

Spinning side bands are reported in the Figure as symbols. Milled Ca(BH$_4$)$_2$ + MgH$_2$ composite presents two sharp lines at -30 and -32 ppm belonging to the boron atoms within the [BH$_4$]$^-$ anion. Since the starting material is composed of the two polymorphs γ and β, with different boron coordination, every peak corresponds to a different structure. The signal at -30 ppm belongs to the low temperature phase γ (orthorhombic as the α-phase), while the one at -32 ppm represents the β-Ca(BH$_4$)$_2$ (tetragonal).[119] CaB$_{12}$H$_{12}$ shows a strong signal at -15.4 ppm in accord with the chemical shift already reported in literature for [B$_{12}$H$_{12}$]$^{2-}$ species (-15.6 ppm).[90, 115] MgB$_2$ shows a very pronounced peak at around 100 ppm.[120] The CaB$_6$ spectrum exhibits two lines, at +12 and +0.75 ppm, because of the two different boron sites in its structure. Three broad signals are visible in the spectrum of the desorbed material: +16, -15.6 and -30 ppm. By comparison of the spectrum of the pure reference compounds and that of the desorbed material, the peaks at +16 and at -30 ppm can be assigned to CaB$_6$ and to residual β-Ca(BH$_4$)$_2$ respectively, whereas the signal at -15.6 ppm belongs to CaB$_{12}$H$_{12}$. The same value (-15.6 ppm) was reported in literature by Hwang et al.[90] for the K$_2$B$_{12}$H$_{12}$ dissolved in water. Furthermore, the aforementioned three signals, for desorbed Ca(BH$_4$)$_2$ with NbF$_5$ and TiF$_4$ additives, were recently presented.[121]

There is a slight difference in the chemical shift of the CaB$_{12}$H$_{12}$ used as reference (-15.4 ppm) and the one observed after hydrogen desorption (1st Des. spectrum; -15.6 ppm). Due to the broader peak in the spectrum of the desorbed sample, the estimation of the chemical shift value was not straightforward. A mathematical fitting of the NMR signal, reported in Figure 3.47, was necessary in order to determine the exact value (-15.6 ppm). Generally, the peak could broaden due to the disordered nature of the phase structure.[90] Such structural disorder, for the desorbed material, would contribute to explain why CaB$_{12}$H$_{12}$ cannot be detected by X-rays.

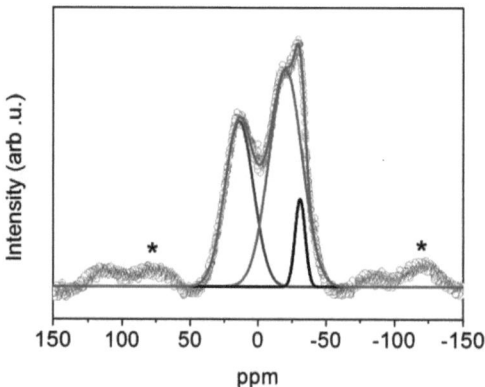

Figure 3.47. Fitting of the ^{11}B{^{1}H} MAS-NMR spectrum at room temperature of milled Ca(BH$_4$)$_2$ + MgH$_2$ composite after desorption at 400 °C in vacuum (1st Des.). Experimental spectrum (green); CaB$_6$ (blue); CaB$_{12}$H$_{12}$ (red); β-Ca(BH$_4$)$_2$ (black). Side bands are indicated by ✶.

The observed NMR spectrum of the milled Ca(BH$_4$)$_2$ + MgH$_2$ composite desorbed at 400 °C in vacuum can be fitted very accurately using three peaks at the peak positions of CaB$_6$, CaB$_{12}$H$_{12}$ and residual β-Ca(BH$_4$)$_2$ signal.

NMR measurements of the materials after first (re)absorption reaction are reported in Figure 3.48.

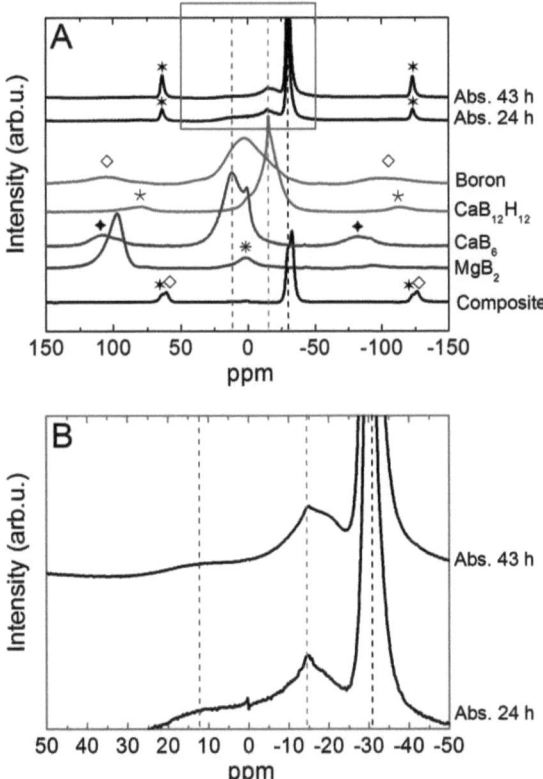

Figure 3.48. A: $^{11}B\{^1H\}$ MAS-NMR spectrum at room temperature of milled $Ca(BH_4)_2$ + MgH_2 composite (re)absorbed at 350 °C and 145 bar H_2 for 24 (Abs. 24 h) and 43 hours (Abs. 43 h). Side bands are indicated by ✷, ◊, ✱, ♦, ✶. B: inset of the spectra corresponding to the (re)absorbed materials.

Two signals are visible in both spectra of the (re)absorbed materials. The signal at -30 ppm corresponds to an orthorhombic polymorph of the $Ca(BH_4)_2$. XRD in Figure 3.44 confirms it to be α-$Ca(BH_4)_2$. The signal around -15 ppm belongs to the $CaB_{12}H_{12}$ phase. It is emphasised in the inset reported in Figure B. In addition, Figure 3.48 B shows the lower intensity for the signal corresponding to the CaB_6 phase at 43 hours respect to that at 24 hours.

Figure 3.49 reports the NMR spectra for the materials after first, second and third hydrogen desorption reaction.

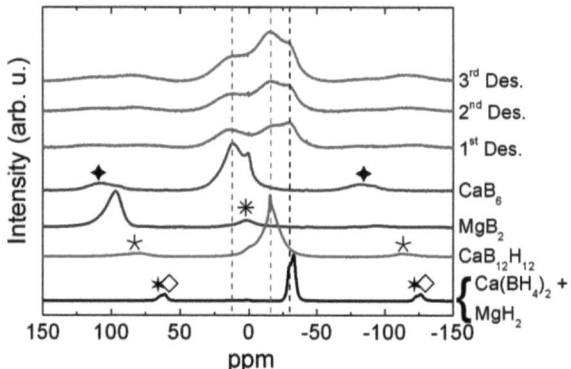

Figure 3.49. $^{11}B\{^1H\}$ MAS-NMR spectra at room temperature of $Ca(BH_4)_2$ + MgH_2 composite after first (1st Des.), second (2nd Des.) and third (3rd Des.) hydrogen desorption at 400 °C in vacuum. Side bands are indicated by ✶, ◊, ★, ✻ and ◆.

The same $^{11}B\{^1H\}$ MAS-NMR spectrum is observed for all the desorbed materials. All of them show three signals at +16, -15.6 and -30 ppm belonging to CaB_6, $CaB_{12}H_{12}$ and residual β-$Ca(BH_4)_2$ respectively. XRD analysis (Fig. 3.40) reveals that pure Mg and CaH_2 are also present among the final products. Figure 3.49 highlights the increase of the relative intensity of the $CaB_{12}H_{12}$ signal and therefore its quantity, during cycling compared to both signals of CaB_6 and residual β-$Ca(BH_4)_2$.

3.5.6 Is the formation of the $Ca_4Mg_3H_{14}$ phase a necessary reaction step during the decomposition of the $Ca(BH_4)_2$ + MgH_2 composite ?

It was already showed in section 3.5.3 that the formation of $Ca_4Mg_3H_{14}$ occurs during the decomposition of $Ca(BH_4)_2$ + MgH_2 composite. The ternary Ca-Mg-H phase forms because CaH_2, produced by the (de)hydrogenation of $Ca(BH_4)_2$, reacts with MgH_2. Later on, the ternary phase decomposes leading to the formation of Mg, CaH_2 and H_2. In order to understand whether the formation of the $Ca_4Mg_3H_{14}$ phase is really a necessary reaction step, the same composite material was prepared by modifying the previous preparation procedure. The former analyses were carried out with the material ball milled for 5 hours in a stainless steel vial using stainless steel spheres. In the present case, $Ca(BH_4)_2$ and MgH_2 powder were simply mixed together for 5 minutes inside the vial without further help of milling tools. Afterwards, Differential Scanning Calorimetry and *in-situ* XRD analyses were performed on the non-milled composite material. DSC analysis is reported in Figure 3.50.

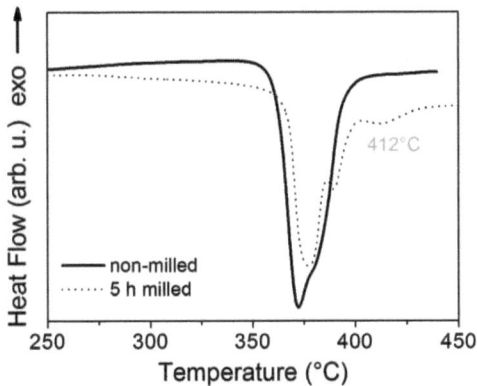

Figure 3.50. DSC curve for the 5 hours ball milled (dotted) and for the non-milled (solid) $Ca(BH_4)_2$ + MgH_2 composite at 50 ml min^{-1} argon flow.

At first sight, the curves look rather similar with the two endothermic peaks within the temperature range of 360-375 °C. The endothermic signals correspond to the $Ca(BH_4)_2$ and MgH_2 hydrogen desorption reaction. The main difference between the two calorimetric curves reported in Figure 3.50 is the absence of the endothermic peak at 412 °C for the non-milled material. This signal corresponds to the hydrogen desorption reaction of the $Ca_4Mg_3H_{14}$ phase. The absence of the signal at 412 °C implies that the formation of Ca-Mg-H ternary phase does not occur.

The *in-situ* XRD analysis for the non-milled $Ca(BH_4)_2$ + MgH_2 composite is reported in Figure 3.51

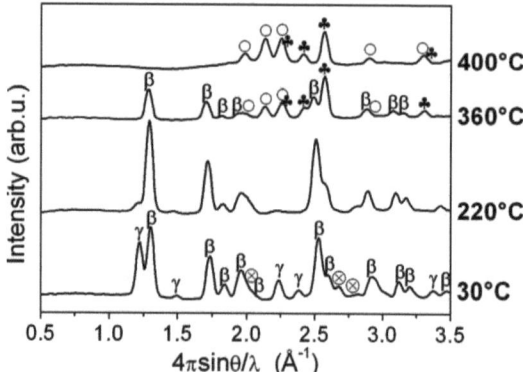

Figure 3.51 SR-PXD patterns of non-milled Ca(BH$_4$)$_2$ + MgH$_2$ composite. The experiment was carried out by heating in vacuum from RT up to 400 °C with 3 °C min^{-1} as constant heating rate. γ-Ca(BH$_4$)$_2$ (γ); β-Ca(BH$_4$)$_2$ (β); MgH$_2$ (⊗); Mg (♣); CaH$_2$ (O). The measurement was performed at the synchrotron Hasylab, DESY (Hamburg), at the beamline D3.

The diffraction data at 30 °C presents the peaks of the γ-, β-Ca(BH$_4$)$_2$ and the MgH$_2$ phase. Their relative abundance is 29 (± 5 % error), 56 (± 5 % error) and 15 wt. % (± 5 % error) respectively. In the temperature range of 30-220 °C the low temperature γ-Ca(BH$_4$)$_2$ gradually transforms into the high temperature β-Ca(BH$_4$)$_2$ phase. In the XRD pattern at 360 °C the peaks belonging to the Mg and CaH$_2$ phase are visible together with the reflections of the β-Ca(BH$_4$)$_2$ phase. At this temperature, the Bragg peaks of MgH$_2$ are already disappeared indicating that its hydrogen desorption reaction has already ended. The decrease of the intensity of the reflections belonging to β-Ca(BH$_4$)$_2$ and the increase of those corresponding to Mg and CaH$_2$ indicates that a hydrogen desorption reaction is occurring. The results so far presented would therefore suggest that MgH$_2$ desorbs before β-Ca(BH$_4$)$_2$. At 400 °C all the β-Ca(BH$_4$)$_2$ has desorbed and the final products are visible: CaH$_2$ and Mg. At 400 °C MgB$_2$ is not visible as well as other boron-phases (e.g. boron, CaB$_6$, CaB$_{12}$H$_{12}$) implying their amorphous or nanocrystalline status as already observed in the case of the transition metal fluorides doped Ca(BH$_4$)$_2$ and of the milled Ca(BH$_4$)$_2$ + MgH$_2$ composite system. ^{11}B{^{1}H} MAS-NMR experiments are necessary in order to detect those amorphous or nanocrystalline phases but it was not possible to carry them out.

The *in-situ* XRD analysis confirms that the formation of the $Ca_4Mg_3H_{14}$ phase is not a necessary reaction step but probably a side reaction. A change in the material preparation procedure (no milling) contributed to anticipate the MgH_2 desorption with respect to the $Ca(BH_4)_2$ avoiding the formation of the ternary Ca-Mg-H phase. This effect is likely linked to the modification of the microstructure.

3.5.7 Is the hydrogen back pressure influencing the decomposition path of the $Ca(BH_4)_2$ + MgH_2 composite system?

In order to obtain a more general understanding of the decomposition paths of the $Ca(BH_4)_2$ + MgH_2 composite system, several experiments, at different temperatures and pressures, were performed. The decomposition reaction was studied at 400 °C (in vacuum and 1 bar H_2 pressure) and at 350 °C (in vacuum, 1 bar and 5 bar H_2 pressure).

Thermodynamic calculations (thermodynamic database and first principle calculations) performed by Kim *et al.*[48] at 350 °C predicts that under 1 bar of hydrogen pressure the (de)hydrogenation of $Ca(BH_4)_2$ + MgH_2 composite, beside CaH_2, would lead to the formation of MgB_2 whereas vacuum would promote the formation of CaB_6 (and Mg).

Figure 3.52 reports the volumetric analysis of the first hydrogen desorption reaction for the $Ca(BH_4)_2$ + MgH_2 composite system at 400 °C both in vacuum and at 1 bar H_2 pressure.

The Figure shows that both samples start desorbing hydrogen at ca. 350 °C. The overall hydrogen capacity (ca. 7 wt. %) is released in 3 hours. The amount of hydrogen delivered in this case is slightly higher compared to that observed in the volumetric curves in Figure 3.38 and 3.45. This is likely related to the absence of side products (e.g. MgO) in the decomposition materials. Figure 3.52 shows that 1 bar hydrogen back pressure does not change the decomposition path of the pure $Ca(BH_4)_2$ + MgH_2 composite system respect to the vacuum atmosphere. The curves coincide.

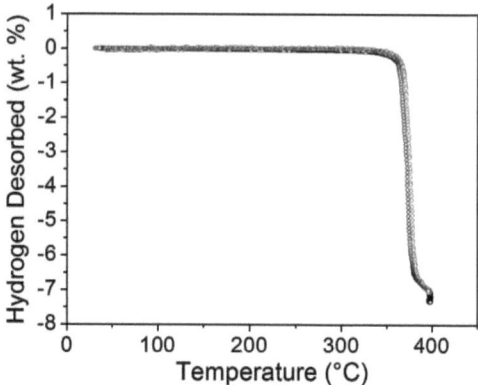

Figure 3.52. Volumetric measurements showing the desorption curves over the temperature. Desorption curve of the Ca(BH$_4$)$_2$ + MgH$_2$ composite in vacuum (black) and at 1 bar H$_2$ pressure (blue). The experiments were carried out by heating the samples from room temperature up to 400 °C.

XRD analysis on the decomposition products, formed in vacuum and in 1 bar H$_2$ pressure, are reported in Figure 3.53 for the Ca(BH$_4$)$_2$ + MgH$_2$ composite system.

Figure 3.53 evidences the Bragg peaks of both the Mg and CaH$_2$ phase. No signals belonging to MgO are visible. This would justify the higher amount of hydrogen delivered during the desorption reaction reported in Figure 3.52 compared to that reported in Figure 3.38. Furthermore, no reflections belonging to any boron-phase are detectable (e.g. CaB$_6$, CaB$_{12}$H$_{12}$ and MgB$_2$). It was already reported that CaB$_6$ cannot be observed in XRD patterns due to its small dimensions.

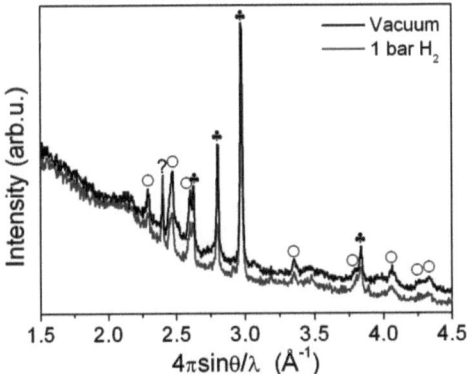

Figure 3.53. XRD spectra of the (de)hydrogenated powders in vacuum and 1 bar H_2 pressure at 400 °C. CaH_2 (O); Mg (♣); ? (artefact). The measurements were performed at the Institute for Metallic Materials at the Leibniz Institute for Solid State and Materials Research (Dresden).

The materials, desorbed at 350 °C in vacuum and at 1 bar H_2 pressure, are reported in Figure 3.54.

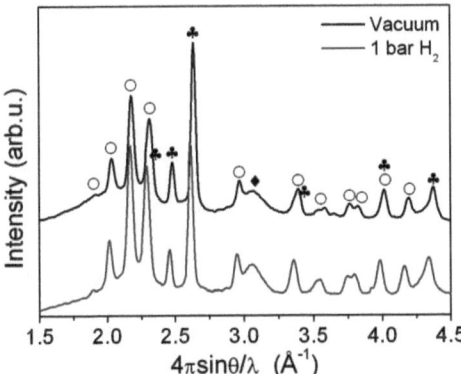

Figure 3.54. XRD spectra of the (de)hydrogenated powders at 350 °C in vacuum and 1 bar H_2 pressure. CaH_2 (O); Mg (♣); MgO (♦). The measurement was performed at the synchrotron Hasylab, DESY (Hamburg), at the beamline D3.

The Bragg peaks shown in Figure 3.54 correspond to the CaH$_2$, Mg and MgO phase. The halo in the scattering vector value range of 1.5-2.5 (Å^{-1}) belongs to the most intense CaB$_6$ peak known to be in nanocrystalline or amorphous-like status. No trace of MgB$_2$ phase is visible.

Due to the fact that, so far, 1 bar hydrogen back pressure did not show any influence on the desorption reaction path of the Ca(BH$_4$)$_2$ + MgH$_2$ composite system, another experiment at 350 °C with 5 bar H$_2$ back pressure was performed. For comparison purposes, the volumetric curve at 5 bar H$_2$ is reported in Figure 3.55 together with that measured in vacuum.

Figure 3.55 shows the slower desorption kinetics for the sample at 5 bar H$_2$ pressure compared to the material measured in vacuum. Although the desorption reaction starts for both samples around 350 °C, with 5 bar H$_2$ pressure, almost 8 hours are necessary to desorb 4.8 wt. % hydrogen. This value represents the 53 % of the theoretical capacity (9.1 wt. % H$_2$) contained within the Ca(BH$_4$)$_2$ + MgH$_2$ composite system. X-ray diffraction on the desorbed materials (shown later) will clarify the reasons for such a lower amount of hydrogen delivered. Instead, the Ca(BH$_4$)$_2$ + MgH$_2$ composite, at 350 °C and in vacuum, desorbs ca. 7 wt. % of hydrogen in 3.5 hours.

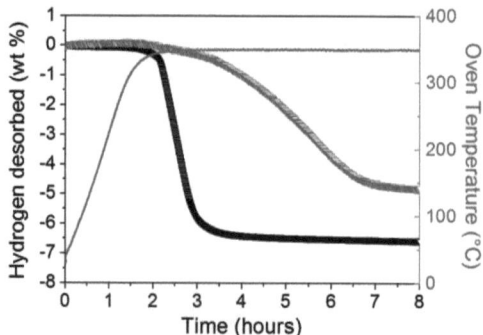

Figure 3.55. Volumetric measurements showing the desorption curves over the time. Temperature (red curve). Desorption curve of the Ca(BH$_4$)$_2$ + MgH$_2$ composite in vacuum (black) and at 5 bar H$_2$ pressure (green). The experiments were carried out by heating the samples from room temperature up to 350 °C.

XRD analysis on the Ca(BH$_4$)$_2$ + MgH$_2$ composite system desorbed at 350 °C and 5 bar H$_2$ pressure is reported in Figure 3.56. For comparison purposes, the XRD pattern of the material desorbed in vacuum is reported as well.

The XRD spectrum of the material desorbed in 5 bar H$_2$ pressure shows the Bragg peaks of the Ca$_4$Mg$_3$H$_{14}$ and of the MgH$_2$ phase. Presence of the ternary Ca-Mg-H phase is explained by the reaction between CaH$_2$ (formed upon (de)hydrogenation of Ca(BH$_4$)$_2$) and MgH$_2$ phase. MgH$_2$ does not desorb at all in these experimental conditions. Presence of CaB$_6$ cannot be observed.

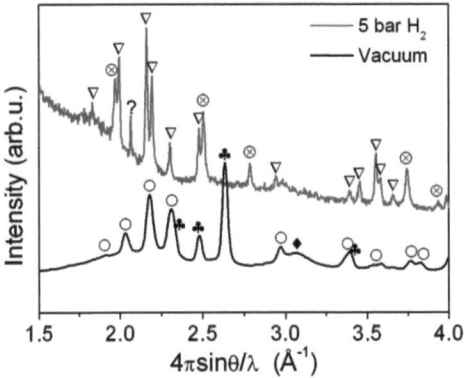

Figure 3.56. XRD pattern of the (de)hydrogenated powders at 350 °C in vacuum and 5 bar H$_2$ pressure. CaH$_2$ (O); Mg (♣); MgO (♦); Ca$_4$Mg$_3$H$_{14}$ (▽); MgH$_2$ (⊗); ? (artefact). The measurements were performed at the Institute for Metallic Materials at the Leibniz Institute for Solid State and Materials Research (Dresden) (spectrum at 5 bar H$_2$) and at the synchrotron MAX-lab, Lund (Sweden) at the beamline I711 (spectrum in vacuum).

The amount of hydrogen released during the desorption reaction at 5 bar H$_2$ pressure (4.8 wt. %) corresponds to ca. 2.4 mol H$_2$. Therefore the (de)hydrogenation path can be approximately represented by the following reaction scheme:

(3) Ca(BH$_4$)$_2$ + x MgH$_2$ → y Ca$_4$Mg$_3$H$_{14}$ + z MgH$_2$ + w CaB$_6$ + 2.4 H$_2$

Reaction 3 does not include the $CaB_{12}H_{12}$ phase that might have formed upon (de)hydrogenation. $^{11}B\{^1H\}$ MAS-NMR analysis were not performed on these samples therefore we cannot confirm or exclude its presence among the decomposition products. However, formation of $CaB_{12}H_{12}$ upon (de)hydrogenation was already observed in the case of the $Ca(BH_4)_2$ + MgH_2 composite desorbed in vacuum conditions.

3.6 Effect of NbF$_5$ and TiF$_4$ on the sorption properties of the Ca(BH$_4$)$_2$ + MgH$_2$ composite system

The addition of NbF$_5$ and TiF$_4$ (shown in section 3.2) has demonstrated to be absolutely necessary to promote the reversible formation of the borohydrides in the single component $Ca(BH_4)_2$ system. The $Ca(BH_4)_2$ + MgH_2 composite system has shown the achievement of partial reversibility upon (re)hydrogenation. As in case of the single component system this seems to be linked to the formation of the $CaB_{12}H_{12}$ phase during the (de)hydrogenation process. Therefore, in order to improve the reversibility of the $Ca(BH_4)_2$ + MgH_2 composite system, 5 mol % of NbF$_5$ or TiF$_4$ were used as additive.

Two different samples were prepared adding 0.05 mol of NbF$_5$ (purity 99%) and of TiF$_4$ (purity 98%), purchased by Alfa Aesar, to the $Ca(BH_4)_2$ + MgH_2 composite system. A description of the samples preparation is provided in the experimental section 2.2.2.

The characterisation of the sorption properties of the $Ca(BH_4)_2$ + MgH_2 composite system milled with additives will be reported in the following sections by means of volumetric measurements, calorimetric techniques, *in* and *ex-situ* XRD analysis, EXAFS and ^{11}B Magic Angle Spinning-Nuclear Magnetic Resonance.

3.6.1 The (De)hydrogenation Reaction

The volumetric analysis of the first hydrogen desorption reaction of the $Ca(BH_4)_2$ + MgH_2 composite system milled with NbF$_5$ and TiF$_4$ additives was performed in a Sievert's type apparatus heating from room temperature (25 °C) up to 400 °C with a constant heating rate of 3 °C min^{-1} under vacuum.

X-ray diffraction patterns for all the materials milled with additives are presented in Figure 3.57.

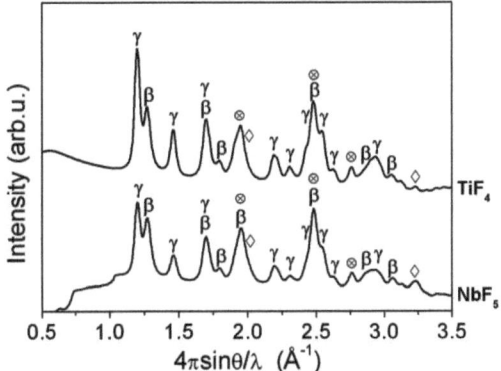

Figure 3.57. XRD patterns after milling of the Ca(BH$_4$)$_2$ + MgH$_2$ composite milled with 0.05 mol of NbF$_5$ and TiF$_4$. γ-Ca(BH$_4$)$_2$ (γ); β-Ca(BH$_4$)$_2$ (β); MgH$_2$ (⊗); CaF$_2$ (◇). The measurement was performed at the synchrotron MAX-lab, Lund (Sweden) at the beamline I711.

All the samples indicate presence of γ-Ca(BH$_4$)$_2$ (space group Pbca, orthorhombic phase)[102], β-Ca(BH$_4$)$_2$ [71], MgH$_2$ and CaF$_2$ with different abundance. The values are reported in Table 3.1.

Phase	Ca(BH$_4$)$_2$ + MgH$_2$ + **NbF$_5$**	Ca(BH$_4$)$_2$ + MgH$_2$ + **TiF$_4$**
γ-Ca(BH$_4$)$_2$	40 wt. % (± 5 error)	41 wt. % (± 5 error)
β-Ca(BH$_4$)$_2$	45 wt. % (± 5 error)	46 wt. % (± 5 error)
MgH$_2$	9 wt. % (± 5 error)	9 wt. % (± 5 error)
CaF$_2$	6 wt. % (± 5 error)	4 wt. % (± 5 error)

Table 3.1

The presence of CaF$_2$ hints to an irreversible reaction between the TM-fluoride and the borohydride already during the milling process. No Bragg peaks belonging to NbF$_5$ or TiF$_4$ phase are visible in both the XRD patterns suggesting a rather fine distribution of the additives.

Figure 3.58 shows the kinetic curves of the first desorption reaction of the composite milled with the additives, obtained by thermovolumetric measurements. For comparison the desorption curve for pure ball milled $Ca(BH_4)_2$ + MgH_2 composite system is reported as well.

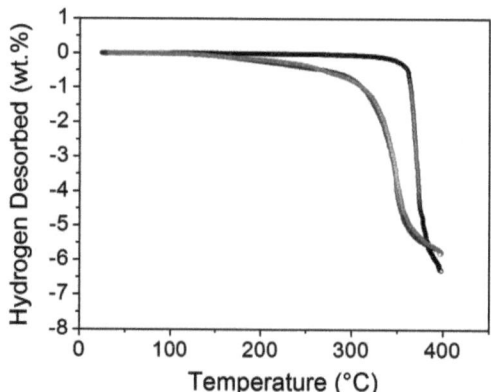

Figure 3.58. Volumetric measurements showing the desorption curves over the temperature. Desorption curve of the pure $Ca(BH_4)_2$ + MgH_2 composite (black); $Ca(BH_4)_2$ + MgH_2 + (0.05 mol) NbF_5 (blue); $Ca(BH_4)_2$ + MgH_2 + (0.05 mol) TiF_4 (dark yellow). Experiments were carried out by heating the samples from room temperature up to 400 °C in static vacuum (starting value 0.02 bar).

Figure 3.58 evidences that the addition of TM-fluorides changes the desorption kinetics: in the ball milled $Ca(BH_4)_2$ + MgH_2 composite the hydrogen starts to be released around 350 °C while in the samples with additives begins already in the range of 125-225 °C. Such a low hydrogen desorption temperature reaction was already observed for the $Ca(BH_4)_2$ system milled with transition metal fluorides additives (section 3.2). As visible in Figure 3.58, the material doped with NbF_5 and TiF_4 desorbs at markedly lower temperatures than pristine $Ca(BH_4)_2$ + MgH_2.

While the pure ball milled $Ca(BH_4)_2$ + MgH_2 desorbs 6.4 wt. % hydrogen in 3 hours under the applied conditions, the samples with additives desorb less hydrogen because of the formation of some side products like CaO (shown later). After 3 hours, $Ca(BH_4)_2$ + MgH_2 doped with NbF_5 and TiF_4 desorbs ca. 6.2 wt. % and ca. 5.8 wt. % H_2 respectively.

Now the key question is whether these additives, are able to improve the reversibility with respect to the ball milled $Ca(BH_4)_2$ + MgH_2 composite system.

3.6.2 Thermal Analysis

As Figure 3.59 shows, the endothermic events, in case of the samples milled with the TM-fluoride additives, are shifted to lower temperatures respect to the pure $Ca(BH_4)_2$ + MgH_2 composite.

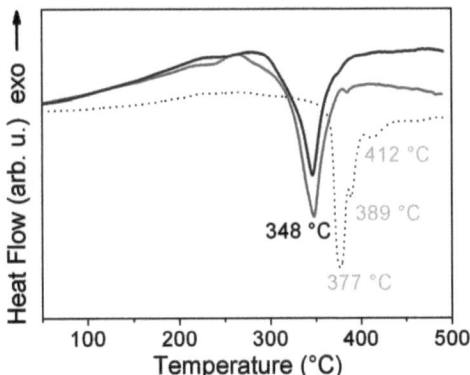

Figure 3.59. DSC curves at 50 ml min^{-1} argon flow of ball milled $Ca(BH_4)_2$ + MgH_2 (dotted); $Ca(BH_4)_2$ + MgH_2 + NbF_5 (blue); $Ca(BH_4)_2$ + MgH_2 + TiF_4 (dark yellow).

The curves belonging to the samples milled with NbF_5 and TiF_4 show that the hydrogen desorption step takes place at significantly lower temperature than in case of the ball milled $Ca(BH_4)_2$ + MgH_2 composite system. In particular, with both NbF_5 and TiF_4, the desorption reaction starts below 300 °C. Note that, in the case of the doped samples, the peak temperature is also well below that observed for the pure ball milled composite system. Furthermore, Figure 3.59 evidences that, apart from the hydrogen desorption peak, the samples milled with additives do not present any other endothermic signals. However, this might be due to the rather broad shape of the endothermic signals which enclose other events.

3.6.3 The (Re)hydrogenation Reaction

The desorption products were subsequently (re)absorbed at 145 bar H_2 pressure and 350 °C for 24 hours. Due to the low quality of the volumetric absorption curves, they will not be reported here. In order to obtain quantitative information about the amount of hydrogen reversibly absorbed, subsequent desorption tests on the (re)absorbed powders were performed. XRD diffraction was performed on the (re)absorbed materials doped with NbF_5 and TiF_4 and the results are reported in Figure 3.60.

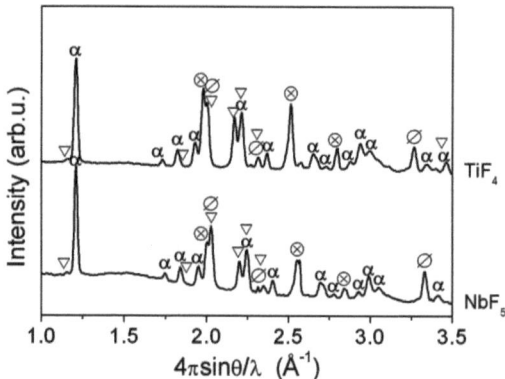

Figure 3.60. XRD of $Ca(BH_4)_2$ + MgH_2 doped with NbF_5 and TiF_4 after (re)absorption reaction at 350 °C and 145 bar H_2 pressure for 24 hours. α-$Ca(BH_4)_2$ (α); $CaF_{2-x}H_x$ (Ø); $Ca_4Mg_3H_{14}$ (∇); MgH_2 (⊗). The measurements were performed at the synchrotron MAX-lab, Lund (Sweden) at the beamline I711.

Figure 3.60 displays, for both the patterns, the reflections of the α-$Ca(BH_4)_2$ phase. Hence, partial reversible formation was achieved. The (de)hydrogenated product $CaF_{2-x}H_x$ is also visible. Compared to the pure system, no trace of CaH_2 can be detected in the XRD patterns. Instead, reflections of the $Ca_4Mg_3H_{14}$ are visible. CaH_2 is not present because it has reacted with MgH_2 to form the ternary Ca-Mg-H phase. As already observed for the $Ca(BH_4)_2$ system, the (re)hydrogenated material shows only the α- modification whereas the starting $Ca(BH_4)_2$ powder contained both the γ- and β- polymorph. Therefore this is another confirmation of the higher energetic stability of the α- modification respect to the others polymorphs.

Figure 3.61 shows the thermovolumetric measurements corresponding to the second hydrogen desorption reaction for the samples milled with TiF$_4$ and NbF$_5$. As shown by the Figure, the sample with TiF$_4$ desorbs 3.2 wt. % whereas the sample with NbF$_5$ desorbs 4 wt. % hydrogen, corresponding to 55 % and 65 % reversibility, respectively. However, the kinetics is hampered compared to the first desorption process as well as to the 2nd desorption of the pure composite. Section 3.5.4 showed that the pure system offers 55 % reversibility upon first hydrogen desorption. This is exactly the same value observed in the case of the sample doped with TiF$_4$. The sample with NbF$_5$ exhibits a slightly higher reversibility (65 %) compared to the pure composite system but still far away from the desired value. Thus, under the applied conditions the addition of the transition metal fluorides additives does not improve any further the performance already showed by the pure Ca(BH$_4$)$_2$ + MgH$_2$ composite system.

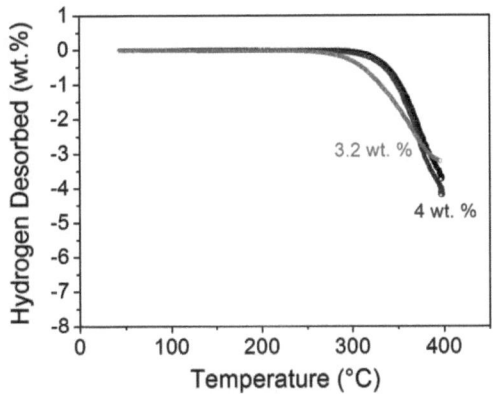

Figure 3.61. Volumetric measurements showing the (re)absorption curves over the temperature. (Re)absorption curve of the Ca(BH$_4$)$_2$ + MgH$_2$ composite (black); Ca(BH$_4$)$_2$ + MgH$_2$ + (0.05 mol) NbF$_5$ (blue); Ca(BH$_4$)$_2$ + MgH$_2$ + (0.05 mol) TiF$_4$ (dark yellow). Experiments were carried out by heating the samples from room temperature up to 350 °C at 145 bar H$_2$ for 24 hours.

3.6.4   Ca(BH$_4$)$_2$ + MgH$_2$ + NbF$_5$: *in-situ* Synchrotron Radiation Powder X-ray Diffraction

In-situ SR-PXD was employed to obtain a comprehensive understanding of the sequence of reactions occurring during hydrogen desorption for the samples doped with NbF$_5$ and TiF$_4$.

The *in-situ* SR-PXD patterns over the temperature are reported in Figure 3.62 for the material milled with NbF$_5$.

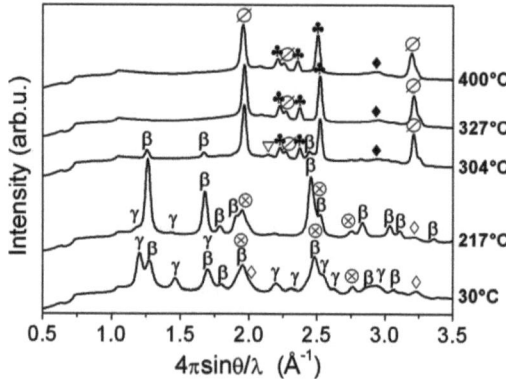

Figure 3.62. SR-PXD patterns of Ca(BH$_4$)$_2$ + MgH$_2$ milled with NbF$_5$. The experiment was carried out by heating the sample in vacuum from RT up to 400 °C with 5 °C min^{-1} as heating rate. γ-Ca(BH$_4$)$_2$ (γ); β-Ca(BH$_4$)$_2$ (β); CaF$_2$ (◇); MgH$_2$ (⊗); CaF$_{2-x}$H$_x$ (∅); Ca$_4$Mg$_3$H$_{14}$ (∇); Mg (♣); MgO (♦). The measurement was performed at the synchrotron MAX-lab, Lund (Sweden) at the beamline I711.

The desorption was studied under static vacuum by heating from room temperature up to 400 °C. Rietveld refinement of the pattern collected at 30 °C indicates that the initial powder is composed of 40 wt. % (± 5 error) low temperature polymorph γ-Ca(BH$_4$)$_2$, 45 wt. % (± 5 error) high temperature β-Ca(BH$_4$)$_2$, 9 wt. % (± 5 error) MgH$_2$ and 6 wt. % CaF$_2$. The latter one must have been formed by an irreversible reaction between Ca(BH$_4$)$_2$ and NbF$_5$ already during milling. This might partly explain why the total amount of hydrogen desorbed from the samples with additives is slightly lower compared to the pure composite system. The pattern measured at 217 °C shows that most of all the γ-phase has transformed into the β-phase. The diffractogram at 304 °C evidences the low intensity of the Bragg peaks belonging to β-Ca(BH$_4$)$_2$ and the presence of Ca$_4$Mg$_3$H$_{14}$, CaF$_{2-x}$H$_x$, Mg and MgO phase. The absence of MgH$_2$ together with the reduced fraction of β-Ca(BH$_4$)$_2$ phase indicate that the hydrogen desorption step, at 304 °C, is approaching the end. The ternary Ca-F-H phase is formed as decomposition product by the reaction between CaH$_2$ and CaF$_2$, as already reported in section

3.2.3 (reaction 1 and 2). The ternary $Ca_4Mg_3H_{14}$ phase (reaction 1), already present at 304 °C, decomposes further, in the temperature range of 304-327 °C, in CaH_2, Mg and H_2 (reaction 2). Yet, pattern at 304 °C displays the signal of the MgO phase. Even though the samples were handled in inert atmosphere, formation of MgO could not be avoided. As already reported by Riktor et al.[72], the oxygen could either derive from impurities in the as-synthesised $Ca(BH_4)_2$ material or from the atmosphere due to a leak in the setup used for the measurement. The pattern at 400 °C shows as final (de)hydrogenated products: $CaF_{2-x}H_x$, Mg and MgO. No trace of boron-phase was detected by in-situ SR-PXD. In order to observe possible amorphous or nanocrystalline B-phase formed, $^{11}B\{^1H\}$ MAS-NMR measurements of the (de)hydrogenated products were performed and the results are reported in the subsection 3.6.8.

3.6.5 $Ca(BH_4)_2$ + MgH_2 + TiF_4: *in-situ* Synchrotron Radiation Powder X-ray Diffraction

The *in-situ* SR-PXD patterns are reported in Figure 3.63 for the material milled with TiF_4.

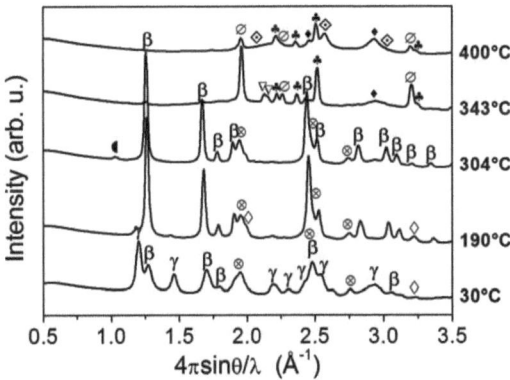

Figure 3.63. SR-PXD patterns of $Ca(BH_4)_2$ + MgH_2 milled with TiF_4. The experiment was carried out by heating in vacuum from RT up to 400 °C with 5 °C min^{-1} as heating rate. γ-$Ca(BH_4)_2$ (γ); β-$Ca(BH_4)_2$ (β); CaF_2 (◇); $Ca_3(^{11}BH_4)_3(^{11}BO_3)$ (●); $CaF_{2-x}H_x$ (∅); $Ca_4Mg_3H_{14}$ (∇); Mg (♣); MgO (♦); CaB_6 (⬦). The measurement was performed at the synchrotron MAX-lab, Lund (Sweden) at the beamline I711.

The desorption reaction was studied under static vacuum by heating from room temperature up to 400 °C. Rietveld refinement of the pattern collected at 30 °C indicates that the initial powder is composed of 41 wt. % (± 5 error) low temperature polymorph γ-Ca(BH$_4$)$_2$, 46 wt. % (± 5 error) high temperature β-Ca(BH$_4$)$_2$, 9 wt. % (± 5 error) MgH$_2$ and 4 wt. % CaF$_2$. The latter one, as aforementioned, must have been formed by an irreversible reaction between Ca(BH$_4$)$_2$ and TiF$_4$ already during milling. The pattern measured at 190 °C shows that almost all the γ-phase has completely transformed into the β-phase. At 304 °C the reflections of the Ca$_3$(^{11}BH$_4$)$_3$(^{11}BO$_3$)[72] phase appear. This phase would be the responsible of the second hydrogen desorption step of the Ca(BH$_4$)$_2$ which was not observed in the case of the sample milled with NbF$_5$. The pattern at 343 °C shows already the decomposition products. The main hydrogen desorption therefore occurs in the temperature range of 304-343 °C with some residual being present in the ternary Ca$_4$Mg$_3$H$_{14}$ phase (pattern at 343 °C). The spectrum at 343 °C evidences the reflections of the ternary Ca$_4$Mg$_3$H$_{14}$ phase (reaction 1) which decomposes further releasing the residual hydrogen, in the temperature range of 343-400 °C, in CaH$_2$, Mg and H$_2$ (reaction 2). Furthermore, at 343 °C, Bragg peaks of the CaF$_{2-x}$H$_x$ phase are visible. Yet, pattern at 343 °C displays the signal of the MgO phase. Even though the samples were handled in inert atmosphere, formation of MgO could not be avoided. As already reported by Riktor *et al.*[72], the oxygen could either derive from impurities in the as-synthesised material or from the atmosphere due to a leak in the setup. The pattern at 400 °C shows as final (de)hydrogenated products: CaF$_{2-x}$H$_x$, Mg, MgO and CaB$_6$. The Bragg peaks of the calcium hexaboride are broad indicating an amorphous-like or nano-size microstructure. These observations were already reported in the case of the pure Ca(BH$_4$)$_2$ milled with NbF$_5$ and TiF$_4$. No trace of other boron-phase (e.g. CaB$_{12}$H$_{12}$) is detected by *in-situ* SR-PXD.

3.6.6 Ca(BH$_4$)$_2$ + MgH$_2$ + NbF$_5$: X-ray Absorption Near Edge Structure

The addition of NbF$_5$ to the Ca(BH$_4$)$_2$ + MgH$_2$ composite system during milling has led to a slight improve of the reversibility (65 %) compared to the pure composite (55 %). As observed for the Ca(BH$_4$)$_2$ system, the NbF$_5$ additive was found to be more effective under the applied conditions than TiF$_4$. In the Ca(BH$_4$)$_2$ + MgH$_2$ composite the addition of TiF$_4$ led to 55 % reversibility. Therefore there is no improvement in respect to the pure composite.

However, both the reaction mechanism and the influence of the additive are not understood yet. The nature of the Nb-phase as well as its oxidation state during sorption reactions was determined by XANES (X-ray Absorption Near Edge Structure). XANES data measured at the Nb K-edge (18986 eV) are shown in Figure 3.64. The Figure includes the spectrum of the material after ball milling and after first hydrogen desorption.

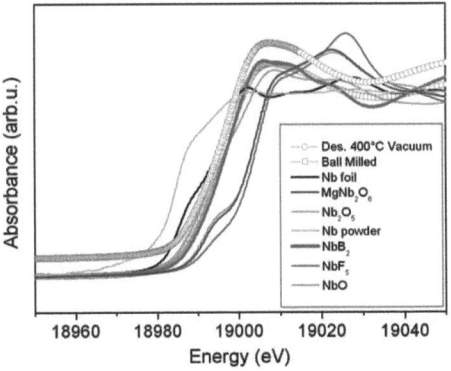

Figure 3.64. XANES data at the Nb K-edge for Ca(BH$_4$)$_2$ + MgH$_2$ + NbF$_5$ (with 5 mol % NbF$_5$) material after ball milling and first hydrogen desorption. The measurements were performed at the synchrotron Hasylab, DESY (Hamburg), at the beamline C.

A direct comparison among the curve of the pure NbF$_5$ compound and those of the material after milling and after hydrogen desorption indicates that a reaction, involving the additive itself and the Ca(BH$_4$)$_2$ + MgH$_2$ composite, has occurred during the milling process. Furthermore, Figure 3.64 indicates that the Nb absorption K-edge does not change anymore after milling process. The curve for the material after hydrogen desorption presents the same absorption K-edge. Hence the oxidation state of the Nb species reduces irreversibly. Figure 3.64 shows that the curves belonging to the milled/desorbed materials fairly match that of the pure NbB$_2$, measured as a reference compound. This would indicate a reduction of the oxidation state of the Nb species from (V) to (II). Furthermore, the similar profile for the curves belonging to the milled/desorbed materials and for the NbB$_2$ along the EXAFS (Extended X-ray Absorption Fine Structure) region (at higher energies respect to the Nb absorption K-edge) would suggest a Nb-B bond in the first coordination shell. This assumption needs to be confirmed by the Fourier Transform analysis of the EXAFS data. Note that *ex-situ* and *in-situ* XRD, in Figure 3.57, 3.60 and 3.62, do not evidence any trace of NbB$_2$ phase. As already reported in literature and in this study, for the Ca(BH$_4$)$_2$ + NbF$_5$

system, the dimensions of such metal boride particles are within a few nanometers only and hence out of the detection range of X-ray diffraction techniques. As visible in Figure 3.63, in the pattern at 400 °C, the CaB_6 phase has an amorphous-like profile. Although the dimensions of this phase are much bigger than those of NbB_2, X-ray diffraction already evidences limit of detection. It is therefore necessary to adopt Transmission Electron Microscopy (TEM) in order to observe and determine the size of such nanoparticles. TEM analysis on the milled and cycled materials were not performed because the addition of transition metal fluoride additives on the $Ca(BH_4)_2 + MgH_2$ system did not evidence any striking improvement respect to the pure composite. However, the formation of metal boride nanoparticles upon milling or hydrogen desorption reaction, in the case of doped systems, is more than a certainty and it is consistent with several works recently reported in literature.

3.6.7 $Ca(BH_4)_2$ + MgH_2 + TiF_4: XANES (X-ray Absorption Near Edge Structure)

The addition of TiF_4 to the $Ca(BH_4)_2 + MgH_2$ composite system does not present any striking effect on the reversibility. A value of 55 % reversibility was observed in the pure $Ca(BH_4)_2 + MgH_2$ composite system as well as in the one doped with TiF_4. However, for the understanding of the reaction mechanism it is necessary to know the nature of the Ti phase formed upon hydrogen sorption process. X-ray diffraction cannot help any further concerning this point because, generally, the dimensions of these phases are within the nanometer range. XANES represents a powerful tool on this regard because it allows determining the oxidation state of the transition metals.

XANES analysis at the Ti K-edge (4966 eV), including the curve for the ball milled and (re)absorbed material, are reported in Figure 3.65.

A comparison, in the pre-edge region of the spectra, among the curve of the ball milled material, TiF_4, TiO_2 and Ti_2O_3, suggests that a reaction, involving the additive itself and the $Ca(BH_4)_2 + MgH_2$ composite, has occurred during the milling process. In fact, the pre-edge at 4971 eV for the TiF_4 curve has disappeared in the case of the milled sample. The reaction during the milling process leads to the formation of CaF_2 as shown by XRD in Figure 3.57. The Ti absorption K-edge, in the case of the milled sample, matches the value observed for TiO_2 and Ti_2O_3. TiF_4 is not taken into account because it has reacted with $Ca(BH_4)_2$ to form CaF_2 (Figure 3.57). Oxide formation might come from contaminations due to the high sensitivity of alkali tetrahydroborates to moisture.[27] In addition, Riktor et al.[72] reported the starting Ca(BH$_4$)$_2$ material containing already solvent traces or oxygen-containing impurities. Nevertheless, within the rising edge region (at higher energies compared to the pre-edge region) the curve of the milled sample overlaps that of pure TiO$_2$. This result is similar to what observed for the TiF$_4$ doped Ca(BH$_4$)$_2$ system. Experiments performed by Buslaev et al.[114] would exclude TiO$_2$ as a product of the hydrolysis reaction of TiF$_4$. The combination of the observations reported for the milled sample, within the rising edge region and at the absorption edge would suggests that Ti has an oxidation state between IV (TiO$_2$) and III (Ti$_2$O$_3$).

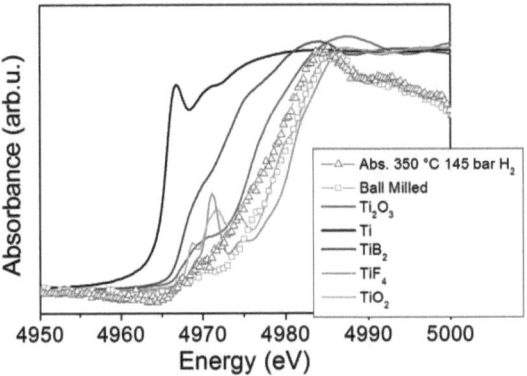

Figure 3.65. XANES data at the Ti K-edge for the Ca(BH$_4$)$_2$ + MgH$_2$ + TiF$_4$ (with 5 mol % TiF$_4$) material after ball milling and (re)absorption process. The measurements were performed at the synchrotron Hasylab, DESY (Hamburg), at the beamline A1.

The (re)absorbed sample shows further reduction in the oxidation state of Ti towards the absorption K-edge value observed for TiB$_2$. Instead, within the rising edge region, the curve of the (re)absorbed sample falls between those of pure Ti$_2$O$_3$ and TiO$_2$. The combination of these observations suggests either that Ti has an oxidation state between III (Ti$_2$O$_3$) and II (TiB$_2$) or a mixture of both different chemical states. The same behaviour was already reported by Deprez et al.[103] in case of Ti-isopropoxide doped LiBH$_4$-MgH$_2$ RHC. Unfortunately, the results for the (re)absorbed material could not reported here due to the low quality of the curve.

TEM analysis on the milled and cycled materials were not performed because the addition of transition metal fluoride additives on the Ca(BH$_4$)$_2$ + MgH$_2$ system did not evidence any striking improvement respect to the pure composite.

3.6.8 ^{11}B{^{1}H} Magic Angle Spinning–Nuclear Magnetic Resonance

The detection of boron-phase by XRD has demonstrated to be tangled therefore the materials after first desorption, after subsequent (re)absorption and after second hydrogen desorption reaction were analysed by ^{11}B{^{1}H} MAS-NMR. The NMR spectra of the pure and doped desorbed composite systems are shown in Figure 3.66 together with those of selected reference compounds.

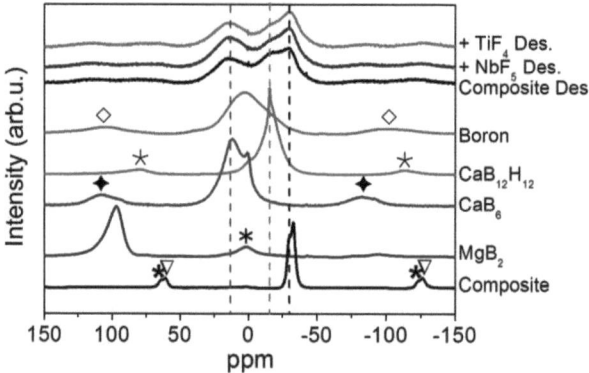

Figure 3.66. ^{11}B{^{1}H} MAS-NMR spectra at room temperature of the pure Ca(BH$_4$)$_2$ + MgH$_2$ composite system and doped with NbF$_5$ and TiF$_4$ after first hydrogen desorption at 400 °C in vacuum. Side bands are indicated by ★, ▽, ✱, ♦, ✶, ◇.

Spinning side bands are reported in the Figure as symbols. Milled Ca(BH$_4$)$_2$ + MgH$_2$ pure composite presents two sharp lines at -30 and -32 ppm belonging to the boron atoms within the [BH$_4$]$^-$ anion. Since the starting material is composed of the two polymorphs γ and β, with different boron coordination, every peak corresponds to a different structure. The signal at -30 ppm belongs to the low temperature phase γ (orthorhombic as the α-phase), while the one at -32 ppm represents the β-Ca(BH$_4$)$_2$ (tetragonal).[119] CaB$_{12}$H$_{12}$ shows a strong signal at -15.4 ppm according to the chemical shift already reported in literature for [B$_{12}$H$_{12}$]$^{2-}$ species (-15.6

ppm).[90, 115] MgB$_2$ shows a very pronounced peak around 100 ppm.[120] The CaB$_6$ spectrum exhibits two lines, at +12 and +0.75 ppm, because of the two different boron sites in its structure.[91] The same three broad signals are visible in the spectra of the pure and doped desorbed materials: +16, -15.6 and -30 ppm.[121] The peaks at +16 and at -30 ppm correspond to the CaB$_6$ and to the residual β-Ca(BH$_4$)$_2$ respectively. The signal at -15.6 ppm belongs to CaB$_{12}$H$_{12}$. The same value (-15.6 ppm) was reported in literature by Hwang *et al.*[90] for K$_2$B$_{12}$H$_{12}$ dissolved in water. Furthermore, the aforementioned three signals, for desorbed Ca(BH$_4$)$_2$ with NbF$_5$ and TiF$_4$ additives, were recently presented. [121]. In the case of the composite materials doped with NbF$_5$ and TiF$_4$, the amount of CaB$_{12}$H$_{12}$ formed (respect to the signal of CaB$_6$) is slightly lower compared to the pure composite system. This can be observed in Figure 3.66. The spectra of the (re)absorbed materials, reported in Figure 3.67, evidence two signals. The peak at -30 ppm corresponds to an orthorhombic polymorph of the Ca(BH$_4$)$_2$. XRDs in Figure 3.60 confirm that it is α-Ca(BH$_4$)$_2$. The signal around -15 ppm belongs to the CaB$_{12}$H$_{12}$ phase. In the chemical shift region where CaB$_6$ signal falls, a low intensity peak is visible probably due to residual not reacted calcium hexaboride. In principle, it should have reacted in total, upon (re)hydrogenation reaction, to form α-Ca(BH$_4$)$_2$.

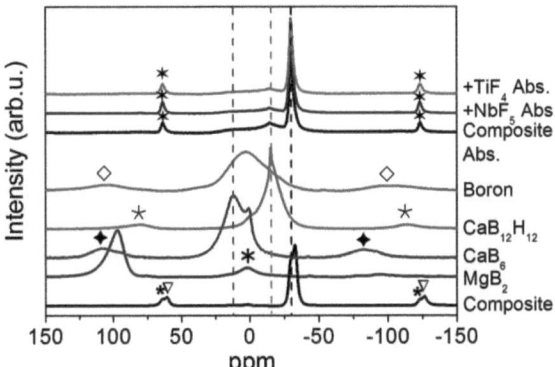

Figure 3.67. ^{11}B{^{1}H} MAS-NMR spectrum at room temperature of pure Ca(BH$_4$)$_2$ + MgH$_2$ composite system and doped with NbF$_5$ and TiF$_4$ after subsequent (re)absorption at 350 °C and 145 bar H$_2$ for 24 h. Side bands are indicated by ✶, ◊, ✱, ♦, ✯, ◇.

The $^{11}B\{^1H\}$ MAS-NMR spectra for the materials after second hydrogen desorption are reported in Figure 3.68.

The same three signals at +16, -15.6 and -30 ppm, observed for the samples after first hydrogen desorption, are visible in the spectra of the materials after second hydrogen desorption. The signals belong to CaB_6, $CaB_{12}H_{12}$ and residual β-$Ca(BH_4)_2$ respectively. XRD analysis (Figure 3.62 and 3.63) reveals that pure Mg and $CaF_{2-x}H_x$ are also present among the final products. Traces of MgO are visible as well. Figure 3.68 shows that only for the TiF_4 doped composite system, the amount of $CaB_{12}H_{12}$ is either reduced or its particles size become smaller upon cycling. If the first result is likely, it should, in principle, be confirmed by a better reversibility compared to the pure composite system. Instead, a value of 55 % is observed for the TiF_4 doped system which is exactly the same calculated for the pure $Ca(BH_4)_2$ + MgH_2 composite.

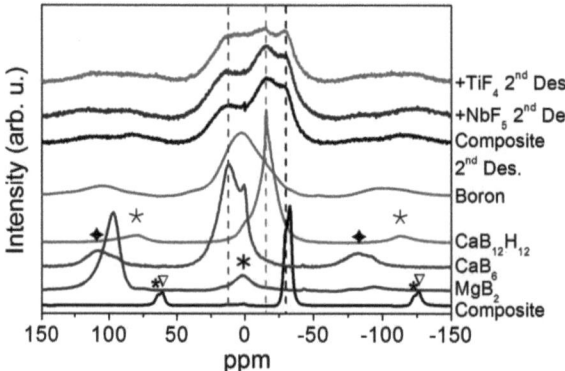

Figure 3.68. $^{11}B\{^1H\}$ MAS-NMR spectra at room temperature of the $Ca(BH_4)_2$ + MgH_2 composite system doped with NbF_5 and TiF_4 after second hydrogen desorption at 400 °C in vacuum. Side bands are indicated by ★, ▽, ✲, ◆, ⋆.

4 Discussion

In this section, the sorption properties of the $Ca(BH_4)_2$ and $Ca(BH_4)_2 + MgH_2$ composite system as well as the role of the additives on their reversible formation are discussed referring to the results presented in chapter 3. Concerning the $Ca(BH_4)_2$ system, transition metal fluorides (NbF_5 and TiF_4) revealed to have a fundamental role on both its reaction kinetics and reversible formation. Regarding the $Ca(BH_4)_2 + MgH_2$ composite system, instead, Mg (formed after desorption) exhibits a stronger effect compared to the addition of transition metal fluorides (NbF_5 and TiF_4).
In this chapter, the effect of the additives referring to the interplanar misfit between phases at the interface will be discussed.

4.1 $Ca(BH_4)_2$ system

The identification of the desorbed materials is fundamental for the understanding of the (re)absorption mechanism. Several decomposition pathways are reported in literature for pure $Ca(BH_4)_2$ and they are listed in Table 4.1 with the corresponding calculated thermodynamic data.

Reaction	ΔH	ΔS	T	Hydrogen capacity
$Ca(BH_4)_2 \leftrightarrow 2/3\ CaH_2 + 1/3\ CaB_6 + 10/3\ H_2$	*32*[61] 37.04[66] 40.8[63] 40.6[48]	108.51 111.3 109.3	68 94 98[62]	9.6
$Ca(BH_4)_2 \rightarrow CaH_2 + 2\ B + 3\ H_2$	57.3[48] *71.56*[66]	105.7	263	8.7
$Ca(BH_4)_2 \rightarrow CaB_2H_2 + 3\ H_2$	*68.51*[66]			8.7
$Ca(BH_4)_2 \rightarrow CaB_2H_4 + 2\ H_2$	*77.35*[66]			5.8
$Ca(BH_4)_2 \rightarrow CaB_2H_6 + H_2$	31.09[66]	83.19	100	2.9
$Ca(BH_4)_2 \rightarrow 1/6\ CaB_{12}H_{12} + 5/6\ CaH_2 + 13/6\ H_2$	39.2[63] 35.8- 37.9[65] 31.34[66]	106.5 92.50	99 66	6.3

Table 4.1. Decomposition pathways for $Ca(BH_4)_2$. ΔH: reaction enthalpy (kJ mol^{-1} H_2). ΔS: reaction entropy (J K^{-1} mol^{-1} H_2). T: calculated hydrogen desorption temperature at p = 1 bar H_2 pressure. Values calculated at 0 K are indicated in *italic*.

Table 4.1 shows the boron compounds (CaB_6, B (boron), CaB_2H_2, CaB_2H_4, CaB_2H_6 and $CaB_{12}H_{12}$) which can be formed upon decomposition of $Ca(BH_4)_2$. $[B_{12}H_{12}]^{2-}$ containing species were predicted to be likely during decomposition of tetrahydroborates and their chemical stability is known to be rather high.[67] However, their detection is difficult due to their amorphous status. Wang et al.[65] predicted the existence of several polymorphs of $CaB_{12}H_{12}$ during (de)hydrogenation reaction of $Ca(BH_4)_2$. Riktor et al.[68] and Lee et al.[69] observed the formation of CaB_2H_x (x = 2) and of CaB_mH_n phase respectively. Zhang et al.[66] found the phase proposed by Riktor et al.[68] too unstable to be a decomposition product (see Table 4.1). DFT and PEGS (Prototype Electrostatic Ground-State) calculations performed by Zhang et al.[66] suggested the CaB_2H_6 phase to be more likely. The structures for CaB_2H_6 and $CaB_{12}H_{12}$ are reported in Figure 4.1.

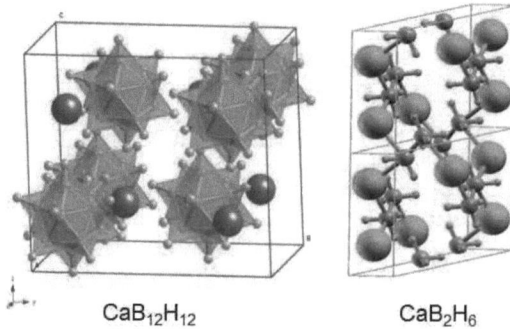

Figure 4.1. Predicted crystal structures of $CaB_{12}H_{12}$ (reproduced from Ref. 63) and CaB_2H_6 (reproduced from Ref. 66).

$^{11}B\{^1H\}$ MAS-NMR, on $Ca(BH_4)_2$ after hydrogen desorption reaction, evidenced abundant amount of amorphous boron (Fig. 3.6). A small shoulder at +16 ppm hints to the formation of a small quantity of CaB_6 as well. Figure 3.6 does not evidence formation of $CaB_{12}H_{12}$ or other CaB_2H_x phase. But can we truly exclude formation of such phases? The calculated enthalpies reported in Table 4.1, indicate that CaB_2H_x compounds may be formed upon (de)hydrogenation. A tight competition for the calculated reaction enthalpy values of CaB_2H_6 and $CaB_{12}H_{12}$ (31.09 and 31.34 kJ mol^{-1} H_2 respectively) is observed. Instead, reactions leading to CaB_2H_2 and to CaB_2H_4 exhibit reaction enthalpy values well above those calculated for CaB_2H_6 and $CaB_{12}H_{12}$ (see Table 4.1). Hence, their formation is less likely to occur during decomposition of Ca(BH$_4$)$_2$. Unfortunately, so far, there are no ^{11}B{^{1}H} MAS-NMR chemical shifts data reported on the CaB$_2$H$_2$ and CaB$_2$H$_6$ phases probably because their chemical synthesis is not straightforward. However, their crystal structure is rather different compared to that of CaB$_{12}$H$_{12}$ hinting to others ^{11}B{^{1}H} MAS-NMR chemical shift values. Zhang et al.[66], for instance, report the [B$_2$H$_6$]$^{2-}$ anion being calculated referring to an ethane-like structure.

In the end, formation of CaB$_2$H$_6$ and/or CaB$_{12}$H$_{12}$ cannot be entirely ruled out since the broad peak of the boron phase (Fig. 3.6) might hide others low intensity peaks over a wide chemical shift range. Besides, simultaneous presence of boron and CaB$_6$ in the desorbed materials (Fig. 3.6) would suggest that Ca(BH$_4$)$_2$ follows simultaneously two different decomposition paths. Given the thermodynamic database provided by Kim et al.[48], reported in Table 4.1, the reaction leading to boron ($\Delta H°$=57.3 kJ mol^{-1} H$_2$; $\Delta S°$=105.7 J K^{-1} mol^{-1} H$_2$) evidences a free energy of $\Delta G°$=-8.6 kJ mol^{-1} H$_2$ at 350 °C. The reaction leading to CaB$_6$ ($\Delta H°$=40.6 kJ mol^{-1} H$_2$; $\Delta S°$=109.3 J K^{-1} mol^{-1} H$_2$) provides a free energy of $\Delta G°$=-27.5 kJ mol^{-1} H$_2$ at 350 °C. Thermodynamically, the reaction path ending in CaB$_6$ is more favourable. Therefore, kinetic effects must have more important role under the experimental conditions reported in this work.

XRD, after (re)hydrogenation at 130 bar H$_2$ and 350 °C for 24 hours (Fig. 3.5) does not show any trace of calcium borohydride. Instead, the same decomposition products (CaO, CaH$_2$ and boron (by NMR)) are still visible. The unsuccessful reversible formation of Ca(BH$_4$)$_2$ can be ascribed to the presence of boron formed upon desorption. Formation of boron is undesirable because of its known reluctance to react.[44] Its inertness might be related to the strong covalent B-B bond in the structure (B(s)→B(g): 560 kJ mol^{-1}).[92] Orimo et al.[43] reported LiBH$_4$ to be reversibly formed only at 600 °C and 350 bar H$_2$ starting from LiH and boron. Mauron et al.[122], reduced the conditions of (re)hydrogenation for the same desorption products to 600 °C and 155 bar H$_2$ pressure. Kim et al.[48] report formation of Ca(BH$_4$)$_2$ being negligible even after keeping the decomposition products (CaH$_2$, CaB$_6$ and CaO) at 350 °C under 90 bar H$_2$ for 48 hours. Long-range diffusion of CaH$_2$ and CaB$_6$ is believed to be the main barrier for the (re)hydrogenation reaction.[123].Ronnebro et al.[73] were able to synthesise Ca(BH$_4$)$_2$, with a yield of 60 %, from a mixture of CaH$_2$ and CaB$_6$ with Pd and TiCl$_3$ by applying 700 bar H$_2$ and temperatures of 400-440 °C.

The works mentioned above evidence how the formation of boron can suppress the reversible formation of [BH$_4$]$^-$. Supplement of additives in the starting mixture containing CaB$_6$ (e.g.

Ronnebro et al.[73]) seems to be fundamental to overcome the intrinsic high kinetic energy barrier for the formation of tetrahydroborates.

4.2 Role of additives on the $Ca(BH_4)_2$ system

Section 3.2 evidenced the effectiveness of the addition of TiF_4 and NbF_5 on the reversible formation of $Ca(BH_4)_2$. Instead, supplement of VF_3, VF_4 and TiF_3 demonstrated to be ineffective. Formation of boron was observed by $^{11}B\{^1H\}$ MAS-NMR after hydrogen desorption of the samples with VF_3, VF_4 and TiF_3 (Figure 3.20). On the contrary, formation of CaB_6 was detected in the TiF_4 and NbF_5 doped $Ca(BH_4)_2$ material after hydrogen desorption (Figure 3.20). XANES and TEM analysis (section 3.2.6, 3.2.7 and 3.2.9), on TiF_4 and NbF_5 doped $Ca(BH_4)_2$ material, revealed irreversible formation of TiB_2 and NbB_2 nanoparticles (in the range of 10-20 nm) upon desorption reaction. In case of TiF_4 doped material after hydrogen desorption, the Ti-compound exists as a mixture of Ti_2O_3 and TiB_2 (Figure 3.19). However, formation of transition metal borides is in agreement with some studies already reported in literature in case of transition metal chlorides and isopropoxides additives.[103, 111, 112] In addition, the TEM picture in Figure 3.22 (d) highlights the nanoparticles in the desorbed sample. Formation of NbB_2 nanoparticles, in case of NbF_5 additive, is believed to occur already during the milling process (Figure 3.64, NbF_5 doped composite system) due to the low melting temperature of the additive (m. p. 79 °C). In fact temperatures of ca. 100 °C can be reached within a vial during milling.[105] This would increase the reactivity of NbF_5 as demonstrated by studies on the composition of its vapour pressure at these temperatures.[124]

The first set of kinetic measurements on $Ca(BH_4)_2$ milled with VF_3, VF_4 and TiF_3 (Fig. 3.9) and leading to boron after (de)hydrogenation, evidences an incubation period upon desorption. TiF_4 and NbF_5 doped $Ca(BH_4)_2$ materials which lead to the formation of CaB_6 during hydrogen desorption, do not show such a plateau. XRD measurements performed on samples prior (de)hydrogenated to the concentration range of the plateau (Fig. 3.11) evidence the same phases for all the samples ($Ca_3(^{11}BH_4)_3(^{11}BO_3)$ and $CaF_{2-x}H_x$) therefore the plateau phenomenon is believed being caused by the formation of an intermediate, amorphous or nanocrystalline phase as observed in other systems.[103, 111, 112] The formation of the intermediate, amorphous or nanocrystalline phase would implicate that by addition of TiF_3, VF_3 and VF_4 the reaction path is altered if compared to pure non-milled, TiF_4 and NbF_5 doped $Ca(BH_4)_2$ system, where such an incubation period is not observed. It has to be mentioned, however, that the incubation period, in case of the TiF_3, VF_3 and VF_4 additives, was not reproduced in the second set of kinetic measurements (Fig. 3.10) although samples preparation was performed following the same procedure.

In case of $LiBH_4$-MgH_2 composite system, Bösenberg et al.[112] showed that transition metal borides act as heterogeneous nucleation sites for the formation of MgB_2. In addition, they avoid the incubation period upon desorption and, at the same time, act as a microstructure refiner improving the sorption kinetics. In our study, the incubation period was not observed upon desorption of the TiF_4 and NbF_5 doped $Ca(BH_4)_2$ systems (Fig. 3.9) and CaB_6 was detected after decomposition (Fig. 3.20). The formation of TiB_2 and NbB_2 nanoparticles during desorption could have contributed to suppress the incubation period observed in case of TiF_3, VF_3 and VF_4 additives, favouring the formation of CaB_6. In addition, such nanoparticles could have supported the refinement of the microstructure improving consequently the sorption kinetics.

Yet, Bösenberg et al.[111], in case of Zr-isopropoxide addition in the $LiBH_4$-MgH_2 composite system, reports that ZrB_2 nanoparticles are located in the interfaces and grain boundaries, while the MgB_2 grains are embedded in a LiH matrix in the desorbed state. Transition-metal borides and MgB_2 have the same crystal structure (hexagonal) therefore the borides provide coherent interfaces for nucleation, regardless of crystallographic planes and ledges, favouring heterogeneous nucleation of MgB_2. In our study, TEM images in dark field mode coupled with Selected Area Electron Diffraction (Figure 3.21 and 3.22) detect the nanoparticles and show their good distribution within the matrix. However, the pictures cannot say whether the nanoparticles lie in the grain boundaries. To provide such information, high resolution TEM should be carried out. However, the rapid decomposition of the powder under the electron beam would not make it banal. Nevertheless, the localisation of the transition metal boride nanoparticles in the grain boundaries might be sound since sorption reactions involve considerable mass transport which could have moved the nanoparticles to the interfaces.

The following part discusses the ability of transition metal borides nanoparticles to support heterogeneous nucleation of CaB_6, considered to play a key role for the reversible formation of $Ca(BH_4)_2$, and their role as a grain refiner during sorption reactions. Transition-metal borides (hexagonal) and CaB_6 (cubic) have a different crystal structure. However, the borides could still provide interfaces with low interfacial energy supporting heterogeneous nucleation of CaB_6.

A fundamental requirement for heterogeneous nucleation is a low interfacial energy. Other necessary requirements for an efficient heterogeneous nucleation are a good distribution of the nucleation agents as well as a sufficient amount of it. Generally, chemical contributions are considered playing a secondary role.

Across an interface, the maximum probability to observe atom row matching in consecutive atom rows is maximised if the planes which contain the atom rows in the two phases have very similar interplanar d-spacings and are arranged to meet edge to edge in the interface.[125] The planes normally considered are the close-packed or nearly close-packed planes. A close-packed plane corresponds, in a XRD pattern, to the plane with the highest X-ray diffraction intensity.[125]

The relative difference in the d spacing of any two close-packed or nearly close-packed planes between two phases is called d-value mismatch. If the d-value mismatch is below a critical value (PCV) then this plane pair has potential to form Orientation Relationship (ORs). In our case, a d-critical mismatch value below or equal 6 % is considered reasonable.[125]

There is no rigorous approach to calculate the PCV. Estimated values can be found for known systems in databases. However, for new plane pairs, like in our study, calculations of the d-value mismatch are necessary because no data are available for these interface energies. They were estimated from the interplanar distance of the crystal structures of the crystallographic phases by the following relationship:

$$\delta = \frac{a1-a2}{0.5(a1+a2)}$$

where δ corresponds to the d-value mismatch (%), *a1* is the interplanar distance of the precipitated phase, *a2* is the interplanar distance of the matrix phase.

The calculated d-value mismatches between possible matching planes in several systems are reported in Table 4.2. Determination of the lattice misfit for the $CaF_{2-x}H_x$ phase was not possible due to its unknown stoichiometry. In addition to the transition-metal borides formed by the additives, CaH_2, CaF_2 as well as compounds forming by possible oxygen contamination such as CaO and MgO are evaluated.

Table 4.2 reports the d-value mismatches calculated respect to the $\{111\}_{CaB6}$. CaB_6 has a cubic lattice structure (spatial group Pm-3m) and 4.145 Å as lattice parameter.

The {200} CaO plane provides the lowest d-value mismatch (0.14). This value is however misleading because both our results (section 3.1.3, 3.1.4 and 3.1.5) and those presented by Kim *et al.*[48] demonstrate that no reversible formation of $Ca(BH_4)_2$ can be achieved when

CaB$_6$ and CaO (and CaH$_2$ of course) have to react together to calcium borohydride. Kim et al.,[48] report the clear existence of CaB$_6$ (by means of Raman spectroscopy) after desorption reaction performed both at 330 and 480 °C in vacuum. In our case, Fig. 3.5 and 3.6 shows the presence of CaO, CaB$_6$ (besides boron) and CaH$_2$ but no Ca(BH$_4$)$_2$ is obtained after (re)absorption reaction at 350 °C and 130 bar H$_2$ for 24 hours. Distribution constraints of CaO in the matrix might be at the origin of such an inefficient behaviour.

The matching planes were determined respect to the {1011} plane of the TiB$_2$ and NbB$_2$ phase, formed, upon desorption of the Nb- and Ti- Ca(BH$_4$)$_2$ doped system. The {1010} plane of the NbB$_2$ is included as well in Table 4.2 because it refers to its second most intense XRD reflection (95 % of the intensity of the {1011} plane) and therefore the second close-packed plane. TiB$_2$ (a=3.032 Å; c= 3.231 Å) and NbB$_2$ (a=3.09 Å; c= 3.3 Å) have similar lattice parameters and the same hexagonal lattice structure (spatial group P6/mmm). Although the calcium hexaboride structure is different from the one of the transition metal borides, the {111}$_{CaB6}$/{1011}$_{NbB2}$, {111}$_{CaB6}$/{1010}$_{NbB2}$ as well as the {111}$_{CaB6}$/{1011}$_{TiB2}$ plane pairs have the potential to be the matching planes because the d-value mismatch is well below the d-critical mismatch value (6 %). Regarding the results, reported in Table 4.2, a fundamental role of the transition metal boride nanoparticles as supporters for heterogeneous nucleation is confirmed.

Matching planes	d-value mismatch (%)
{111}$_{CaB6}$/{1010}$_{NbB2}$	2.9
{111}$_{CaB6}$/{1011}$_{NbB2}$	2.8
{111}$_{CaB6}$/{1011}$_{TiB2}$	3.15
{111}$_{CaB6}$/{1011}$_{VB2}$	1.9
{111}$_{CaB6}$/{220}$_{CaF2}$	4.1
{111}$_{CaB6}$/{102}$_{CaH2}$	6.2
{111}$_{CaB6}$/{200}$_{CaH2}$	6.2
{111}$_{CaB6}$/{200}$_{CaO}$	0.14

Table 4.2. Calculated d-value mismatch (%) between possible matching plane pairs in several systems.

In case of the Ca(BH$_4$)$_2$ system milled with TiF$_4$ and NbF$_5$, besides the formation of TiB$_2$ and NbB$_2$ nanoparticles, formation of CaF$_2$ is observed due to a reaction between the transition metal fluoride and the borohydride (Figure 3.8). In case of transition metal fluorides doped MgH$_2$ system, Jin et al.[97], could not entirely exclude a supportive role of MgF$_2$ (side product). CaF$_2$ exhibits a d-value mismatch below the critical value therefore it should favour the heterogeneous nucleation of CaB$_6$. Formation of CaB$_6$ is observed in the XRD pattern in Fig. 3.34 and 3.36. However, a critical experiment performed adding CaF$_2$ to Ca(BH$_4$)$_2$ demonstrated no reversible formation of tetrahydroborate after (re)hydrogenation in the same experimental conditions (350 °C and 145 bar H$_2$ for 24 h). Further independent experiments with Ti-isopropoxide and no CaF$_2$ as additives were carried out to confirm the role of the transition metal boride nanoparticles as a heterogeneous nucleation. Ti-isopropoxide represents a fluorine free compound whereas CaF$_2$ cannot lead to the formation of any transition metal boride nanoparticle because it does not contain any transition metal.

The addition of Ti-isopropoxide to the Ca(BH$_4$)$_2$ system has led to the reversible formation of the borohydride upon (re)hydrogenation reaction (Figure 3.27). This behaviour is similar to what observed for the transition metal based additives. Instead, the addition of CaF$_2$ to Ca(BH$_4$)$_2$ did not turn out into the reversible formation of the borohydride (Figure 3.36). Hence, we can confirm that CaF$_2$ does not play any beneficent role in the material mixture but it only exists as a side product. Presence of TiB$_2$ nanoparticles on the Ti-isopropoxide doped Ca(BH$_4$)$_2$ material after (re)hydrogenation reaction, is confirmed by SAED (Figure 3.31). ^{11}B{^{1}H} MAS-NMR shows formation of CaB$_6$ after the desorption reaction (Fig. 3.30). Unfortunately, in our study, there is no analysis confirming the presence of TiB$_2$ nanoparticles formed upon hydrogen desorption reaction in case of Ti-isopropoxide doped sample. However, Deprez et al.[103] reported, for the Ti-isopropoxide doped 2LiBH$_4$-MgH$_2$ composite, that the titanium additive transforms, upon heating, into a mixture of Ti$_2$O$_3$ and TiB$_2$. Simultaneous presence of CaB$_6$ and TiB$_2$ nanoparticles in the desorbed material would confirm the same results obtained for the TiF$_4$ and NbF$_5$ doped Ca(BH$_4$)$_2$ system, i.e. transition metal boride nanoparticles support the heterogeneous nucleation of CaB$_6$ upon desorption reaction, act as a grain refiner improving the sorption kinetics thus leading to the reversible formation of calcium borohydride.

Another result that Ti-isopropoxide doped Ca(BH$_4$)$_2$ system shares with the TiF$_4$ and NbF$_5$ doped materials is a consistent amount of CaB$_{12}$H$_{12}$ observed after decomposition. Instead, no [B$_{12}$H$_{12}$]$^{2-}$ is visible for the TiF$_3$, VF$_3$ and VF$_4$ doped Ca(BH$_4$)$_2$ system. In this case, boron, formed after desorption, exhibits a broad peak over a wide chemical shift range (Fig. 3.6) which might hide others low intensity peaks (CaB$_2$H$_6$ and CaB$_{12}$H$_{12}$). Hence, Ti-isopropoxide, TiF$_4$ and NbF$_5$ doped Ca(BH$_4$)$_2$ materials follow simultaneously two decomposition paths leading to CaB_6 and $CaB_{12}H_{12}$. This is due to the tight competition in the (de)hydrogenation enthalpy values as observed in Table 4.1. We observed that the addition of some transition-metal fluorides (TiF_3, VF_3 and VF_4) led to the formation of boron upon desorption beside no visible trace of $CaB_{12}H_{12}$ (Fig. 3.20). Formation of boron would imply that no transition-metal boride nanoparticles are formed and hence, their absence cannot promote the heterogeneous nucleation of CaB_6. Simultaneous presence of boron and absence of $CaB_{12}H_{12}$ would suggest that formation of $CaB_{12}H_{12}$ upon desorption is driven by the transition-metal boride nanoparticles. From the results reported in this study, it seems that the almost degenerate decomposition pathways (to CaB_6 and $CaB_{12}H_{12}$) are followed when transition-metal boride nanoparticles are present.

4.3 $Ca(BH_4)_2$ + MgH_2 system

Respect to the pure non-milled $Ca(BH_4)_2$ which does not show reversibility upon (re)hydrogenation due to the presence of boron, the $Ca(BH_4)_2$-MgH_2 composite system was found to be reversibly formed without supplement of additives. A reversibility of 55 % was estimated after the first cycle. No transition metal boride nanoparticle could be present within the system prior to desorption, therefore, in order to understand what is the compound promoting the reversibility, XRD and $^{11}B\{^1H\}$ MAS-NMR were performed on the decomposition products. Both the techniques were employed since formation of amorphous and nanocrystalline phases can occur upon (de)hydrogenation. CaH_2, Mg, CaB_6, $CaB_{12}H_{12}$ and MgO were detected to be the materials after decomposition. No formation of MgB_2 was observed. Several decomposition pathways are reported in literature for $Ca(BH_4)_2$ + MgH_2 composite system. They are listed in Table 4.3 with the corresponding calculated thermodynamic data.

Reaction	ΔH	ΔS	T	Hydrogen capacity
$Ca(BH_4)_2 + MgH_2 \leftrightarrow CaH_2 + MgB_2 + 4\ H_2$	46.9[48]	110	151	8.4
$Ca(BH_4)_2 + MgH_2 \leftrightarrow 2/3\ CaH_2 + 1/3\ CaB_6 + Mg + 13/3\ H_2$	45[48]	114.4	120	9.1
$Ca(BH_4)_2 + MgH_2 \leftrightarrow CaH_2 + 2\ B + Mg + 3\ H_2$	57.9[48]	112	238	8.4

Table 4.3. Decomposition pathways for $Ca(BH_4)_2$ + MgH_2. ΔH: standard enthalpy at 25 °C and 1 bar H_2 pressure (kJ mol^{-1} H_2). ΔS: standard entropy at 25 °C and 1 bar H_2 pressure (J K^{-1} mol^{-1} H_2). T: calculated hydrogen desorption temperature at p = 1 bar H_2 pressure.

Table 4.3 evidences a tight competition for the reaction leading to MgB_2 or CaB_6 and Mg. Barkhordarian et al.[46] and Schiavo et al.[126] reported that $Ca(BH_4)_2$ can be synthesised by a mixture of CaH_2 and MgB_2 or CaH_2 and AlB_2 respectively, under hydrogen pressure. The transition metal borides (MgB_2) were reported to have an "unexpected kinetic effect"[46] which enhances the absorption reaction kinetics. The same absorption reaction with boron instead of the boride does not lead to the formation of $Ca(BH_4)_2$ due to kinetic limitations. However, in the present system, upon (de)hydrogenation, MgB_2 or AlB_2 are not reversibly formed.

The calculated d-value mismatches between possible matching planes for several systems in the $Ca(BH_4)_2$ + MgH_2 composite are reported in Table 4.4. The values were determined following the same procedure described in section 4.2.

Matching planes	d-value mismatch (%)
$\{111\}_{CaB6}/\{10\bar{1}1\}_{Mg}$	0.6
$\{111\}_{CaB6}/\{200\}_{MgO}$	2.6
$\{111\}_{CaB6}/\{110\}_{MgH2}$	9
$\{111\}_{CaB6}/\{101\}_{MgH2}$	1.2
$\{111\}_{CaB6}/\{102\}_{CaH2}$	6.2
$\{111\}_{CaB6}/\{200\}_{CaH2}$	6.2

Table 4.4. Calculated d-value mismatch (%) between possible matching plane pairs in several systems

The d-value mismatch reported in Table 4.4, calculated for the $\{111\}_{CaB6}/\{10\bar{1}1\}_{Mg}$ plane pair is the lowest among the other plane pairs (0.6 %). This value is even lower compared to those obtained for the transition metal boride nanoparticles. The 0.6 % value suggests that is Mg the element which support the heterogeneous nucleation of CaB_6 during decomposition of the $Ca(BH_4)_2$ + MgH_2 composite system. This result highlights the "additive-like" role of Mg on the reversible formation of $Ca(BH_4)_2$ at the experimental conditions reported in this work. The d-value mismatch for the $\{111\}_{CaB6}/\{101\}_{MgH2}$ plane pair (however the double the value of Mg) might suggest a significant contribution played by MgH_2 to the heterogeneous nucleation of CaB_6. However, MgH_2, upon desorption, transforms into Mg quickly enough to discard this hypothesis. Moreover, the $\{111\}_{CaB6}/\{200\}_{MgO}$ plane pair observed in Table 4.4 for MgO might imply a supportive role played by the oxide upon (de)hydrogenation reaction. However, both its relative phase abundance after desorption (lower compared to Mg) and its higher d-value mismatch compared to Mg (>4 times higher) suggest for MgO a less important role. Since, usually, nucleation takes place at ledge positions the misfit in the third dimension is also important. CaB_6 crystals can grow in other directions and not only parallel respect to the basal plane of the Mg surface. In addition, the mass transport to the interface is faster when these directions are followed. In our case, the mismatch in the third direction between CaB_6 and Mg is very good (0.2 %). This result strengthens the role of Mg as a supporter of the heterogeneous nucleation of CaB_6 during decomposition of the $Ca(BH_4)_2$ + MgH_2 composite system.

The $Ca(BH_4)_2$-MgH_2 composite system was shown to be partially reversible (55 %). $^{11}B\{^1H\}$ MAS-NMR shows that the $CaB_{12}H_{12}$ phase is formed upon desorption and its phase fraction increases over cycling (Fig. 3.49). This experimental evidence explains why $Ca(BH_4)_2$ suffers limited reversibility. Once $CaB_{12}H_{12}$ is formed upon desorption, due to its high stability in the experimental conditions used in this work [127], it does not participate in the reversible reaction to form $Ca(BH_4)_2$. $^{11}B\{^1H\}$ MAS-NMR analysis, on the (re)absorbed powder (Fig. 3.48) still evidences the $CaB_{12}H_{12}$ signal contributing to explain its reluctance to react. Therefore, during desorption, the $Ca(BH_4)_2$ + MgH_2 composite decomposes into $CaB_{12}H_{12}$, CaH_2, CaB_6, Mg and H_2. Note that $CaB_{12}H_{12}$ is only formed during desorption. Of course, under the experimental conditions reported in this work, the reversible formation will always be partial due to the stable presence of $CaB_{12}H_{12}$. The reaction scheme for the $Ca(BH_4)_2$ + MgH_2 composite is presented in Figure 4.2 in section 4.3.1.

Kim *et al.*[48] predicted by SGTE calculations that, at 350 °C and 1 bar H_2, formation of MgB_2 is likely whereas 350 °C and dynamic vacuum lead to the formation of CaB_6. The XRD presented in this work for the $Ca(BH_4)_2$-MgH_2 composite (Fig. 3.54) show no difference in the decomposition products at 350 °C and 1 bar H_2 or dynamic vacuum. CaH_2, Mg and MgO phase were observed together with a broad peak, in the scattering vector value range of 1.5-2.5 ($Å^{-1}$), which belongs to the most intense CaB_6 signal. The signal broadens due to the nanocrystalline or amorphous-like status. In addition, XRD do not reveal any trace of MgB_2.

Bösenberg *et al.*[128] reported formation of MgB_2 in the $2LiBH_4$-MgH_2 composite system being favoured upon desorption at 400 °C and 5 bar H_2 pressure. The desorption reaction at 350 °C and 5 bar H_2 pressure on $Ca(BH_4)_2$-MgH_2 composite performed in this study, led to the formation of $Ca_4Mg_3H_{14}$ and MgH_2. MgH_2 does not desorb at all in these experimental conditions because 5 bar represents a value higher than its equilibrium pressure (at 350 °C). However, $Ca(BH_4)_2$ desorbs hydrogen leading to the formation of CaH_2 which reacts with

MgH$_2$ to form Ca$_4$Mg$_3$H$_{14}$ (reaction 2 section 3.5.3). It is not possible to observe CaB$_6$ in the spectrum reported in Figure 3.56, due to the most intense peak of Ca$_4$Mg$_3$H$_{14}$ which falls in the same scattering region of CaB$_6$. However, calcium hexaboride was always observed within the decomposition products of the Ca(BH$_4$)$_2$ + MgH$_2$ composite system desorbed at 400 °C in vacuum and at 350 °C in both vacuum and 1 bar H$_2$ pressure. Therefore, it is reliable to assume its presence in the final materials (de)hydrogenated at 350 °C and 5 bar H$_2$ pressure.

4.3.1 Reaction Scheme for the pure milled Ca(BH$_4$)$_2$ + MgH$_2$ composite system

Figure 4.2 reports the general scheme of reaction for the pure milled Ca(BH$_4$)$_2$ + MgH$_2$ composite system.

At 400 °C and vacuum, milled Ca(BH$_4$)$_2$ + MgH$_2$ composite decomposes in CaH$_2$, CaB$_6$, CaB$_{12}$H$_{12}$, Mg and H$_2$. The intermediate step during decomposition is the formation of Ca$_4$Mg$_3$H$_{14}$ which was shown to take place (only in the milled composite material) by reaction between CaH$_2$ and MgH$_2$. This ternary phase decomposes further in CaH$_2$, Mg and H$_2$.

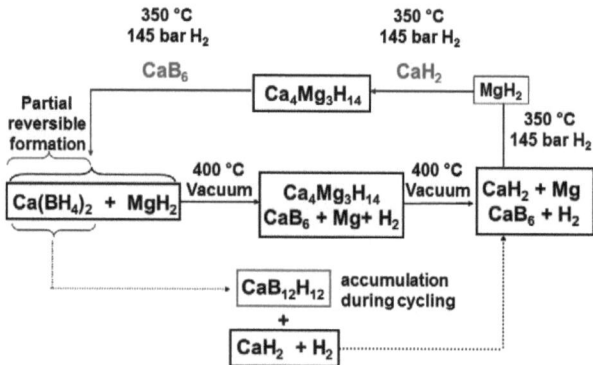

Figure 4.2. Reaction scheme for the desorption and (re)absorption reaction of the pure milled Ca(BH$_4$)$_2$ + MgH$_2$ composite system.

During (re)absorption reaction at 350 °C and 145 bar H_2, Mg (re)hydrogenates to MgH_2 which reacts quickly with CaH_2 to form $Ca_4Mg_3H_{14}$. The ternary phase reacts then with CaB_6 leading to the formation of the $Ca(BH_4)_2$ + MgH_2 composite. In these experimental conditions $CaB_{12}H_{12}$ does not participate to any reaction during (re)absorption process and it accumulates upon further cycling.

4.4 Transition Metal Fluorides doped $Ca(BH_4)_2$ + MgH_2 composite system

Section 3.5.4 showed that the pure $Ca(BH_4)_2$ + MgH_2 composite offers 55 % reversibility upon first (re)hydrogenation reaction. The same value was determined in the case of the TiF_4 doped composite. The sample with NbF_5 exhibits a slightly higher reversibility (65 %) compared to the pure composite but still far away from the desired value.

In the case of the NbF_5 and TiF_4 doped composite material, the amount of $CaB_{12}H_{12}$ formed (respect to CaB_6) is slightly lower compared to the pure composite system (Figure 3.66). This result explains the slight better reversibility observed in case of the NbF_5 doped composite (65 %). We would expect therefore, a higher reversibility for the TiF_4 doped composite system. Instead, the TiF_4 doped and pure $Ca(BH_4)_2$ + MgH_2 system present the same value of reversibility (55 %). Hence, the lower reversibility achieved in case of TiF_4 doped sample could be attributed to side products whose NMR signals fall in the same chemical shift region of the decomposition products. Since the phase fraction of the side products is expected to be low, the respective NMR signals would not be intense enough to overtake those belonging to the decomposition products.

Volumetric measurements confirm again the partial reversibility of the system (Fig. 3.61). $^{11}B\{^1H\}$ MAS-NMR shows the increase of the relative intensity of the $CaB_{12}H_{12}$ signal and therefore its quantity, during cycling compared to both signals of CaB_6 and residual β-$Ca(BH_4)_2$ (Fig. 3.68). This experimental evidence explains the observed limited reversibility for the $Ca(BH_4)_2$ + MgH_2 composite system. Once $CaB_{12}H_{12}$ is formed upon desorption, due to its kinetic stability [127] at these experimental conditions, it does not participate in the reversible reaction to form $Ca(BH_4)_2$. Such behaviour upon (re)hydrogenation can be observed for the $MgB_{12}H_{12}$ as well.[129] $^{11}B\{^1H\}$ MAS-NMR analysis, on the (re)absorbed powder (Figure 3.67) still evidences the $CaB_{12}H_{12}$ signal contributing to explain its reluctance to react. Therefore, during desorption, the $Ca(BH_4)_2$ + MgH_2 doped composite system decomposes into $CaB_{12}H_{12}$, $CaF_{2-x}H_x$, CaB_6, Mg and H_2. Of course, under the experimental conditions reported in this work, the reversible formation will always be partial due to the stable presence of $CaB_{12}H_{12}$.

The addition of transition metal fluorides (NbF_5 and TiF_4) to the $Ca(BH_4)_2 + MgH_2$ composite does not play a fundamental role as, instead, it was observed for the $Ca(BH_4)_2$ system. Formation of transition metal boride nanoparticles upon mechanical or heating process is observed for Ti and Nb doped composite. However, the pure composite was shown to be already reversible without further supplement of additive in the system. This is due to the excellent d-value mismatch at the $\{111\}_{CaB6}/\{10\bar{1}1\}_{Mg}$ interface (0.6 %) and in the third dimension (0.2 %). Transition metal fluorides addition on the composite system might influence the amount of $CaB_{12}H_{12}$ formed upon cycling. However, only in case of NbF_5 addition a slightly better reversibility was observed. The key role in limiting the reversibility is always played by the presence of $CaB_{12}H_{12}$ phase which, in the experimental conditions used in this study, does not decompose.

5 Summary and Outlook

A comprehensive investigation of the effect of selected TM-fluorides on the sorption properties of calcium borohydride was carried out. Calorimetry, FTIR, volumetric measurements, SR-PXD, and Solid State $^{11}B\{^1H\}$ MAS-NMR were employed to study the (de)hydrogenation pathways and characterise the decomposition products. $^{11}B\{^1H\}$ MAS-NMR showed that pure non-milled $Ca(BH_4)_2$ decomposes in CaH_2 and elemental boron. (Re)hydrogenation attempts of the decomposition products did not succeed in the reversible formation of $[BH_4]^-$ at 130 bar H_2 and 350 °C for 24 h. When TM-fluorides (TiF_4 and NbF_5) were added to $Ca(BH_4)_2$, CaF_2 was formed as side product contributing to reduce the hydrogen content of the mixtures. Formation of $CaF_{2-x}H_x$ was detected in the decomposition products due to the reaction between CaH_2 and CaF_2. Elemental boron was formed after desorption of calcium borohydride milled with VF_4, TiF_3, and VF_3. (Re)hydrogenation, under 145 bar H_2 at 350 °C for 20 hours, did not show reversible formation of calcium borohydride. CaB_6 and $CaB_{12}H_{12}$ were formed after hydrogen desorption of $Ca(BH_4)_2$ milled with TiF_4 and NbF_5. CaB_6 and $CaB_{12}H_{12}$ are products of reactions 7 and 11 (section 1.1) that are competing decomposition pathways. Formation of TiB_2 and NbB_2 nanoparticles was also observed after hydrogen desorption. (Re)hydrogenation, under 145 bar H_2 and 350 °C for 20 hours, evidences reversible formation of calcium borohydride. CaB_6 is believed to play a key role in the reversible hydrogenation reaction of $Ca(BH_4)_2$. Its positive function in improving the reversibility compared to elemental boron, known as scarcely reversible, is the reason for the observed effects. The transition metal boride nanoparticles are considered essential to promote heterogeneous nucleation of CaB_6 during hydrogen desorption. The $\{111\}_{CaB6}/\{10\bar{1}1\}_{NbB2}$, $\{111\}_{CaB6}/\{10\bar{1}0\}_{NbB2}$ as well as the $\{111\}_{CaB6}/\{10\bar{1}1\}_{TiB2}$ plane pairs have the potential to be the matching planes because the d-value mismatch is well below the d-critical mismatch value (6 %). Formation of $CaB_{12}H_{12}$ is the reason for the partial reversibility observed at the experimental conditions reported in this study (450 °C, vacuum). The positive effect of the Ti-isopropoxide additive on the reversible formation of $Ca(BH_4)_2$ was shown. A reversibility of 34 % was determined. This value approaches the one observed in case of TiF_4 addition (35 %). Formation of CaB_6 and $CaB_{12}H_{12}$ was observed by NMR after first desorption. SAED reveals presence of TiB_2 nanoparticles in the (re)absorbed material which hints to their supportive role for the heterogeneous nucleation of CaB_6 as already seen in case of TiF_4 and NbF_5 doped $Ca(BH_4)_2$ sample.

This study reports, for the first time, on the experimental confirmation of $CaB_{12}H_{12}$ among the decomposition products of the $Ca(BH_4)_2$ + MgH_2 Reactive Hydride Composite.[130] Yet, evidence of the two simultaneous decomposition paths for $Ca(BH_4)_2$ (leading to both $CaB_{12}H_{12}$ and CaB_6) upon (de)hydrogenation reaction, is showed. CaH_2, Mg, CaB_6, $CaB_{12}H_{12}$ and MgO were formed upon hydrogen desorption reaction. The subsequent (re)hydrogenation reaction led to the partial reversible formation (55 %) of the $Ca(BH_4)_2$ + MgH_2 composite system under 145 bar H_2 at 350 °C for 24 h. Surprisingly, the reversibility was achieved without further addition of additives. The d-value mismatch calculated for the $\{111\}_{CaB6}/\{10\bar{1}1\}_{Mg}$ plane pair is very low (0.6 %). The mismatch in the third dimension is 0.2 %. Mg itself, therefore, it is supposed to be the heterogeneous nucleation agent of CaB_6 during decomposition of the $Ca(BH_4)_2$ + MgH_2 composite system. An assessment of the role of $CaB_{12}H_{12}$ during cycling and its counteractive effect during (re)hydrogenation reaction is performed.

In case of the $LiBH_4$-MgH_2 RHC, Bösenberg[45] reports a mismatch of 3.7 % for the pair (0001)–MgB_2|(0001)–Mg. Therefore, Mg would support heterogeneous nucleation of MgB_2.[45, 112] Nevertheless, in ceramic materials like MgB_2, lattice misfit values greater than 2 % cannot be overcome by elastic deformation. Thus, dislocations must be included during the growth but in this way the interfacial energy would increase. Since nucleation takes place mainly at the ledge positions, the misfit in the third dimension has to be considered. In case of $LiBH_4$-MgH_2 RHC, the misfit of the c-axis between Mg and MgB_2 becomes large (~48 %). Hence, this energy has to be released in dislocations.

In this study, the formation of MgB_2, essence of the Reactive Hydride Composite concept, was never observed, upon hydrogen desorption of the $Ca(BH_4)_2$ + MgH_2 system. At 350 °C, in both vacuum and 1 bar H_2 pressure, CaH_2, CaB_6, Mg and trace of MgO phase were formed. At 400 °C, in both vacuum and 1 bar H_2 pressure, CaH_2 and Mg phase were formed. Presence of CaB_6, in these materials, cannot be confirmed due to its small size which makes it difficult to detect when an high background noise in the XRD data is present (Fig. 3.53). At 350 °C and 5 bar H_2 pressure only $Ca(BH_4)_2$ has decomposed to CaH_2 which has reacted quickly with MgH_2 to form the ternary $Ca_4Mg_3H_{14}$ phase. MgH_2 did not desorb hydrogen in these conditions. The hydrogen back pressure did not lead to the formation of MgB_2 as, on the contrary, was observed for other systems. Nevertheless, it is not excluded that MgB_2 could form in different experimental conditions. This investigation is currently in progress in HZG.

It has to be mentioned that the addition of transition metal fluorides (TiF_4 and NbF_5), to the $Ca(BH_4)_2$ + MgH_2 Reactive Hydride Composite, have shown no successful improvement

(TiF$_4$) or a slightly one (NbF$_5$) on its reversible formation. The two TM-fluoride additives did not even contribute to the suppression of the CaB$_{12}$H$_{12}$ phase formation.

Some strategies aimed to improve the reversible capacity of the Ca(BH$_4$)$_2$ and Ca(BH$_4$)$_2$ + MgH$_2$ composite system are presented hereafter.

The first approach is based on a work[127] recently published which shows that a mixture of CaB$_{12}$H$_{12}$ and CaH$_2$ milled and heated up to 833 K (560 °C) in vacuum, decomposes in CaB$_6$ and H$_2$. Of course, 560 °C can be considered a too high desorption temperature for its technological application as hydrogen storage material. However, if the hydrogen desorption of Ca(BH$_4$)$_2$ is performed at 560 °C in vacuum, no formation of CaB$_{12}$H$_{12}$ should be observed. CaH$_2$ and CaB$_6$ should be the only phases present as decomposition products. Therefore, the (re)hydrogenation reaction, carried out at the same experimental conditions reported in this work (350 °C and 145 bar H$_2$ for 24 h) would lead, in principle, to a fully reversible system stable upon cycling.

The second approach deals with the confinement of borohydrides in porous nanostructures. A recent work shows the reversible formation of NaBH$_4$ (at 325 °C and 60 bar H$_2$) when confined in nanoporous carbon.[131] Bulk NaBH$_4$, upon decomposition, forms Na and Na$_2$B$_{12}$H$_{12}$. In order to favour the (re)hydrogenation reaction, these species have to diffuse and recombine. However, the reaction does not proceed because of their slow diffusion and of the phase segregation. Since nanoconfined NaBH$_4$ shows reversibility, the limited (re)hydrogenation reaction it is not due to the stability of the NaB$_{12}$H$_{12}$ phase formed upon (de)hydrogenation but due to the impossibility of NaH and NaB$_{12}$H$_{12}$ to react with hydrogen to (re)form NaBH$_4$.

The nanoconfinement solves this problem because the entire (de)hydrogenated phases are contained within the pores of the nano-sized matrix (nanoporous carbon). In this way the diffusion distances are greatly reduced allowing the reversible reaction even at milder experimental conditions. This approach might be applied to the Ca(BH$_4$)$_2$ material as well. At 400 or 450 °C in vacuum, CaB$_{12}$H$_{12}$ was observed after decomposition. If the starting Ca(BH$_4$)$_2$ material is impregnated in a nanoporous carbon, the final products CaB$_6$, CaH$_2$ and CaB$_{12}$H$_{12}$ would not be segregated anymore and, due to the reduced mass transport distances, they would react together leading to the reversible formation of Ca(BH$_4$)$_2$. In order to impregnate the Ca(BH$_4$)$_2$ material, the "solution impregnation" (SI) technique has to be used. An aqueous solution of the borohydride has to be prepared and impregnated with the porous carbon material using Schlenk technique. Instead, the "melt infiltration" (MI) cannot be applied to Ca(BH$_4$)$_2$ because clear indications of its melting temperature are not reported. It is rather observed that Ca(BH$_4$)$_2$ decomposes without melting. Such MI technique was successfully applied to MgH$_2$, NaAlH$_4$ and LiBH$_4$.[57]

Formation of M(B$_{12}$H$_{12}$)$_{n/2}$ compounds was observed for NaBH$_4$, LiBH$_4$, Ca(BH$_4$)$_2$ and Mg(BH$_4$)$_2$. The suppression of the formation of such kinetic stable M(B$_{12}$H$_{12}$)$_{n/2}$ compounds could be realised combining the tetrahydroborates with different metal hydrides or metals. In principle, this strategy would change the decomposition pathway with results on both thermodynamics and kinetics. In our specific case (Ca(BH$_4$)$_2$ + MgH$_2$ composite) the combination of the two materials did not succeed in the suppression of CaB$_{12}$H$_{12}$ in respect to the single tetrahydroborate. However, although this study investigates the behaviour of the composite at different temperatures and hydrogen pressures, the ideal conditions for the suppression of the formation of CaB$_{12}$H$_{12}$ still has to be found. Although a significant improvement of the sorption kinetics of Ca(BH$_4$)$_2$ was obtained by the addition of some transition-metal fluorides additives a full reversible capacity could not be achieved. In addition, cycling of the material up to the second desorption, evidence no desirable performance. Limited (re)absorption capacity and continue formation of CaB$_{12}$H$_{12}$ upon each hydrogen desorption cycle would lead, in further cycling, to the accumulation of a too kinetic stable system (CaB$_{12}$H$_{12}$). Concerning the Ca(BH$_4$)$_2$ + MgH$_2$ composite instead, a significant fraction of CaB$_{12}$H$_{12}$ was observed during first desorption reaction whereas much less occurs during the second desorption cycle. This explains why in the third desorption cycle the reversible capacity is equal to 83 %. However, the technological application of the Ca(BH$_4$)$_2$ and Ca(BH$_4$)$_2$ + MgH$_2$ material in combination with a PEM fuel cell is not only hindered by their limited reversible capacity but also because of their slow (re)hydrogenation rates and too high reaction temperatures.

6 Bibliography

1. Agency, 2004.

2. Aleklett K.; Höök M.; Jakobsson K.; Lardelli M.; Snowden S.; Söderbergh B. *The Peak of the Oil Age - Analyzing the world oil production Reference Scenario in World Energy Outlook 2008*. Energy Policy, 2010. **38**, 1398-1414.

3. Schmidt, G. and D. Archer *Climate Change: Too much of a bad thing*. Nature, 2009. **458**, 1117-1118.

4. Monasterky, R. *A Burden Beyond Bearing*. Nature, 2009. **458**, 1091-1094.

5. Züttel, A., Borgschulte A. Schlapbach L. *Hydrogen as a Future Energy Carrier:*. Wiley-VCH. 2008.

6. Min S. K.; Zhang X. B.; Zwiers F. W.; Hegerl G. C. *Human contribution to more-intense precipitation extremes*. Nature, 2011. **470**, 376-379.

7. Pall P.; Aina T.; Stone D. A.; Stott P. A.; Nozawa T.; Hilberts A. G. J.; Lohmann D.; Allen M. R. *Anthropogenic greenhouse gas contribution to flood risk in England and Wales in autumn 2000*. Nature, 2011. **470**, 380-384.

8. http://www.lbst.de/ressources/docs2008/2008-05-21_EWG_Erdoelstudie_D.pdf.

9. Dornheim M., *Thermodynamics of Metal Hydrides: Tailoring Reaction Enthalpies of Hydrogen Storage Materials*. Handbook of Hydrogen Storage, 2010.

10. http://www.hydrogen.energy.gov/pdfs/cryocomp_report.pdf.

11. Schlapbach L.; Zuttel A. *Hydrogen-storage materials for mobile applications*. Nature, 2001. **414**, 353-358.

12. http://www1.eere.energy.gov/hydrogenandfuelcells/storage/storage_challenges.htm.

13. Graham T., *On the absorption and dialytic separation of gases by colloid septa*. Philosophical Transactions of the Royal Society 1866. **156**, 399-439.

14. Libowitz G. G.; Hayes H. F.; Gibb T. R. P. *The System Zirconium–Nickel and Hydrogen*. The Journal of Physical Chemistry, 1958. **62**, 76-79.

15. Yeh X. L.; Samwer K.; Johnson W. L. *Formation of an amorphous metallic hydride by reaction of hydrogen with crystalline intermetallic compounds-A new method of synthesizing metallic glasses*. Applied Physics Letters, 1983. **42**, 242-243.

16. Willems J. J. G.; Buschow K. H. J. *From permanent magnets to rechargeable hydride electrodes*. Journal of the Less Common Metals, 1987. **129**, 13-30.

17. van Vucht J. H. N.; Kuijpers F. A.; Bruning H. C. A. M. *Reversible room-temperature absorption of large quantities of hydrogen by intermetallic compounds*. Philips Research Reports, 1970. **25**, 133-140.

18. Kleperis J.; Wójcik G.; Czerwinski A.; Skowronski J.; Kopczyk M.; Beltowska-Brzezinska M. *Electrochemical behavior of metal hydrides.* Journal of Solid State Electrochemistry, 2001. **5**, 229-249.

19. Reilly J.J.; Wiswall R. H. J. *Alloys for the isolation of hydrogen.* US Patent 3,825,418, 1974.

20. Sandrock G.D., *Development of low cost nickel-rare Earth hydrides for hydrogen storage.* In Hydrogen Storage Systems. W. Seifritz , T.N. Vezeroglu , Editor, 1978, 1625-1656.

21. Reilly J.J.; Wiswall R. H. J. *Formation and properties of iron titanium hydride.* Inorganic Chemistry, 1974. **13**, 218-222.

22. Ovshinsky S. R.; Sapru K.; Dec K.; Hong K. *Hydrogen storage materials and method of making same.* Energy Conversion Devices, 1984.

23. Barkhordarian G.; Klassen T.; Bormann R. *Fast hydrogen sorption kinetics of nanocrystalline Mg using Nb_2O_5 as catalyst.* Scripta Materialia, 2003. **49**, 213-217.

24. Schimmel H. G.; Huot J.; Chapon L. C.; Tichelaar F. D.; Mulder F. M. *Hydrogen Cycling of Niobium and Vanadium Catalyzed Nanostructured Magnesium.* Journal of the American Chemical Society, 2005. **127**, 14348-14354.

25. Bogdanovic, B.; Schwickardi M. *Ti-doped alkali metal aluminium hydrides as potential novel reversible hydrogen storage materials.* Journal of Alloys and Compounds, 1997. **253**, 1-9.

26. Chen P.; Xiong Z. T.; Luo J. Z.; Lin J. Y.; Tan K. L. *Interaction of hydrogen with metal nitrides and imides.* Nature, 2002. **420**, 302-304.

27. Züttel A.; Borgschulte A.; Orimo S.-I. *Tetrahydroborates as new hydrogen storage materials.* Scripta Materialia, 2007. **56**, 823-828.

28. Orimo S. I.; Nakamori Y.; Eliseo J. R.; Zuttel A.; Jensen C. M. *Complex hydrides for hydrogen storage.* Chemical Reviews, 2007. **107**, 4111-4132.

29. Fichtner M., *Nanotechnological aspects in materials for hydrogen storage.* Advanced Engineering Materials, 2005. **7**, 443-455.

30. Bonatto Minella C.; Rongeat C.; Domenech-Ferrer R.; Lindemann I.; Dunsch L.; Sorbie N.; Gregory D. H.; Gutfleisch O. *Synthesis of $LiNH_2$ + LiH by reactive milling of Li_3N.* Faraday Discussions, 2011.

31. Ichikawa T.; Isobe S.; Hanada N.; Fujii H. *Lithium nitride for reversible hydrogen storage.* Journal of Alloys and Compounds, 2004. **365**, 271-276.

32. Chen P.; Xiong Z.; Luo J.; Lin J.; Tan K. L. *Interaction between Lithium Amide and Lithium Hydride.* The Journal of Physical Chemistry B, 2003. **107**, 10967-10970.

33. Gregory D. H. *Lithium nitrides, imides and amides as lightweight, reversible hydrogen stores.* Journal of Materials Chemistry, 2008. **18**, 2321-2330.

34. David W. I. F.; Jones M. O.; Gregory D. H.; Jewell C. M.; Johnson S. R.; Walton A.; Edwards P. P. *A Mechanism for Non-stoichiometry in the Lithium Amide/Lithium Imide Hydrogen Storage Reaction.* Journal of the American Chemical Society, 2007. **129**, 1594-1601.

35. Hino S.; Ichikawa T.; Ogita N.; Udagawa M.; Fujii H. *Quantitative estimation of NH_3 partial pressure in H_2 desorbed from the Li-N-H system by Raman spectroscopy.* Chemical Communications, 2005, 3038-3040.

36. Rajalakshmi N.; Jayanth T.T.; Dhathathreyan K.S. *Effect of Carbon Dioxide and Ammonia on Polymer Electrolyte Membrane Fuel Cell Stack Performance.* Fuel Cells, 2003. **3**, 177-180.

37. Uribe F.A.; Gottesfeld S.; Zawodzinski J.T.A. *Effect of Ammonia as Potential Fuel Impurity on Proton Exchange Membrane Fuel Cell Performance.* Journal of The Electrochemical Society, 2002. **149**, A293-A296.

38. Vajo J. J.; Olson G. L. *Hydrogen storage in destabilised chemical systems.* Scripta Materialia, 2007. **56**, 829-834.

39. Reilly J. J.; Wiswall R. H. *Reaction of hydrogen with alloys of magnesium and nickel and the formation of Mg_2NiH_4.* Inorganic Chemistry, 1968. **7**, 2254-2256.

40. Klassen T.; Oerlich W.; Zeng K.; Bormann R.; Huot J. *Nanocrystalline Mg-based alloys for hydrogen storage* Magnesium Alloys and their Applications, 1998, 308-311.

41. Reilly J. J.; Wiswall R. H. *Reaction of hydrogen with alloys of magnesium and copper.* Inorganic Chemistry, 1967. **6**, 2220-2223.

42. Vajo J. J.; Mertens F.; Ahn C. C.; Bowman R. C.; Fultz B., *Altering Hydrogen Storage Properties by Hydride Destabilisation through Alloy Formation: LiH and MgH_2 Destabilised with Si.* The Journal of Physical Chemistry B, 2004. **108**, 13977-13983.

43. Orimo S.; Nakamori Y.; Kitahara G.; Miwa K.; Ohba N.; Towata S.; Züttel A. *Dehydriding and rehydriding reactions of $LiBH_4$.* Journal of Alloys and Compounds, 2005. **404**, 427-430.

44. Laubengayer A. W.; Hurd D. T.; Newkirk A. E.; Hoard J. L. *Boron. I. Preparation and Properties of Pure Crystalline Boron.* Journal of the American Chemical Society, 1943. **65**, 1924-1931.

45. Bösenberg, U., *$LiBH_4-MgH_2$ Composites for Hydrogen Storage.* PhD Thesis, 2009.

46. Barkhordarian G.; Klassen T.; Dornheim M.; Bormann R. *Unexpected kinetic effect of MgB_2 in reactive hydride composites containing complex borohydrides.* Journal of Alloys and Compounds, 2007. **440**, L18-L21.

47. Barkhordarian, G.; Klassen T.; Bormann R. International Patent, Publication number: WO 2006/063627 A1, 2006.

48. Kim Y.; Reed D.; Lee Y. S.; Lee J. Y.; Shim J. H.; Book D.; Cho Y. W. *Identification of the Dehydrogenated Product of Ca(BH$_4$)$_2$*. Journal of Physical Chemistry C, 2009. **113**, 5865-5871.

49. Miwa K.; Ohba N.; Towata S.; Nakamori Y.; Orimo S. I. *First-principles study on copper-substituted lithium borohydride, (Li$_{1-x}$Cu$_x$)BH$_4$*. Journal of Alloys and Compounds, 2005. **404-406**, 140-143.

50. Brinks H. W.; Fossdal A.; Hauback B. C. *Adjustment of the stability of complex hydrides by anion substitution*. Journal of Physical Chemistry C, 2008. **112**, 5658-5661.

51. Eigen N.; Bosenberg U.; von Colbe J. B.; Jensen T. R.; Cerenius Y.; Dornheim M.; Klassen T.; Bormann R. *Reversible hydrogen storage in NaF-Al composites*. Journal of Alloys and Compounds, 2009. **477**, 76-80.

52. Kang X.-D.; Wang P.; Cheng H.-M. *Advantage of TiF$_3$ over TiCl$_3$ as a dopant precursor to improve the thermodynamic property of Na$_3$AlH$_6$*. Scripta Materialia, 2007. **56**, 361-364.

53. Nakamori Y.; Orimo S. I. *Destabilisation of Li-based complex hydrides*. Journal of Alloys and Compounds, 2004. **370**, 271-275.

54. Yin L.; Wang P.; Fang Z.; Cheng H. *Thermodynamically tuning LiBH$_4$ by fluorine anion doping for hydrogen storage: A density functional study*. Chemical Physics Letters, 2008. **450**, 318-321.

55. Yin L. C.; Wang P.; Kang X. D.; Sun C. H.; Cheng H. M. *Functional anion concept: effect of fluorine anion on hydrogen storage of sodium alanate*. Physical Chemistry Chemical Physics, 2007. **9**, 1499-1502.

56. de Jongh P. E.; Adelhelm P. *Nanosizing and Nanoconfinement: New Strategies Towards Meeting Hydrogen Storage Goals*. ChemSusChem, 2010. **3**, 1332-1348.

57. Nielsen T. K.; Besenbacher F.; Jensen T. R. *Nanoconfined hydrides for energy storage*. Nanoscale, 2011. **3**, 2086-2098.

58. Dornheim M.; Eigen N.; Barkhordarian G.; Klassen T.; Bormann R. *Tailoring Hydrogen Storage Materials Towards Application*. Advanced Engineering Materials, 2006. **8**, 377-385.

59. Gross A. F.; Vajo J. J.; Van Atta S. L.; Olson G. L. *Enhanced Hydrogen Storage Kinetics of LiBH$_4$ in Nanoporous Carbon Scaffolds*. The Journal of Physical Chemistry C, 2008. **112**, 5651-5657.

60. Buchter F.; Lodziana Z.; Remhof A.; Friedrichs O.; Borgschulte A.; Mauron P.; Zuttel A.; Sheptyakov D.; Barkhordarian G.; Bormann R.; Chlopek K.; Fichtner M.; Sorby M.; Riktor M.; Hauback B. C.; Orimo S. I. *Structure of Ca(BD$_4$)$_2$ beta-phase from combined neutron and synchrotron X-ray powder diffraction data and density functional calculations*. Journal of Physical Chemistry B, 2008. **112**, 8042-8048.

61. Miwa K.; Aoki M.; Noritake T.; Ohba N.; Nakamori Y.; Towata S.; Zuttel A.; Orimo S. I. *Thermodynamical stability of calcium borohydride Ca(BH$_4$)$_2$*. Physical Review B, 2006. **74**.

62. Frankcombe T. J. *Calcium Borohydride for Hydrogen Storage: A Computational Study of $Ca(BH_4)_2$ Crystal Structures and the CaB_2H_x Intermediate.* Journal of Physical Chemistry C, 2010. **114**, 9503-9509.

63. Ozolins V.; Majzoub E. H.; Wolverton C. *First-Principles Prediction of Thermodynamically Reversible Hydrogen Storage Reactions in the Li-Mg-Ca-B-H System.* Journal of the American Chemical Society, 2009. **131**, 230-237.

64. Riktor M. D.; Sorby M. H.; Chlopek K.; Fichtner M.; Buchter F.; Züttel A.; Hauback B. C. *In situ synchrotron diffraction studies of phase transitions and thermal decomposition of $Mg(BH_4)_2$ and $Ca(BH_4)_2$.* Journal of Materials Chemistry, 2007. **17**, 4939-4942.

65. Wang L. L.; Graham D. D.; Robertson I. M.; Johnson D. D. *On the Reversibility of Hydrogen-Storage Reactions in $Ca(BH_4)_2$: Characterization via Experiment and Theory.* Journal of Physical Chemistry C, 2009. **113**, 20088-20096.

66. Zhang Y. S.; Majzoub E.; Ozolins V.; Wolverton C. *Theoretical prediction of different decomposition paths for $Ca(BH_4)_2$ and $Mg(BH_4)_2$.* Physical Review B, 2010. **82**.

67. Sivaev I. B.; Sjoberg S.; Bregadze V. I.; Gabel D. *Synthesis of alkoxy derivatives of dodecahydro-closo-dodecaborate anion $B_{12}H_{12}^{2-}$.* Tetrahedron Letters, 1999. **40**, 3451-3454.

68. Riktor M. D.; Sorby M. H.; Chlopek K.; Fichtner M.; Hauback B. C. *The identification of a hitherto unknown intermediate phase CaB_2H_x from decomposition of $Ca(BH_4)_2$.* Journal of Materials Chemistry, 2009. **19**, 2754-2759.

69. Lee J. Y.; Ravnsbaek D.; Lee Y. S.; Kim J.; Cerenius Y.; Shim J. H.; Jensen T. R.; Hur N. H.; Cho Y. W. *Decomposition Reactions and Reversibility of the $LiBH_4$-$Ca(BH_4)_2$ Composite.* Journal of Physical Chemistry C, 2009. **113**, 15080-15086.

70. Kim J.-H.; Jin S.-A.; Shim J.-H.; Cho Y. W. *Thermal decomposition behavior of calcium borohydride $Ca(BH_4)_2$.* Journal of Alloys and Compounds, 2008. **461**, L20-L22.

71. Filinchuk Y.; Ronnebro E.; Chandra D. *Crystal structures and phase transformations in $Ca(BH_4)_2$.* Acta Materialia, 2009. **57**, 732-738.

72. Riktor M. D.; Filinchuk Y.; Vajeeston P.; Bardaji E. G.; Fichtner M.; Fjellvag H.; Sorby M. H.; Hauback B. C. *The crystal structure of the first borohydride borate, $Ca_3(BD_4)_3(BO_3)$.* Journal of Materials Chemistry, 2011. **21**, 7188-7193.

73. Ronnebro E.; Majzoub E. H. *Calcium borohydride for hydrogen storage: Catalysis and reversibility.* Journal of Physical Chemistry B, 2007. **111**, 12045-12047.

74. Rongeat C.; D'Anna V.; Hagemann H.; Borgschulte A.; Züttel A.; Schultz L.; Gutfleisch O. *Effect of additives on the synthesis and reversibility of $Ca(BH_4)_2$.* Journal of Alloys and Compounds, 2010. **493**, 281-287.

75. Kim J. H.; Jin S. A.; Shim J. H.; Cho Y. W. *Reversible hydrogen storage in calcium borohydride $Ca(BH_4)_2$.* Scripta Materialia, 2008. **58**, 481-483.

76. Kim, J. H.; Shim J. H.; Cho Y. W. *On the reversibility of hydrogen storage in Ti- and Nb-catalyzed $Ca(BH_4)_2$.* Journal of Power Sources, 2008. **181**, 140-143.

77. Barkhordarian G.; Jensen T. R.; Doppiu S.; Bösenberg U.; Borgschulte A.; Gremaud R.; Cerenius Y.; Dornheim M.; Klassen T.; Bormann R. *Formation of Ca(BH$_4$)$_2$ from hydrogenation of CaH$_2$+MgB$_2$ composite.* Journal of Physical Chemistry C, 2008. **112**, 2743-2749.

78. Jin S.-A.; Lee Y.-S.; Shim J.-H.; Cho Y. W. *Reversible Hydrogen Storage in LiBH$_4$−MH$_2$ (M = Ce, Ca) Composites.* The Journal of Physical Chemistry C, 2008. **112**, 9520-9524.

79. Fang Z. Z.; Ma L. P.; Kang X. D.; Wang P. J.; Wang P.; Cheng H. M. *In situ formation and rapid decomposition of Ti(BH$_4$)$_3$ by mechanical milling LiBH$_4$ with TiF$_3$.* Applied Physics Letters, 2009. **94**, 044104-3.

80. Geis V.; Guttsche K.; Knapp C.; Scherer H.; Uzun R. *Synthesis and characterization of synthetically useful salts of the weakly-coordinating dianion B$_{12}$Cl$_{12}$$^{2-}$.* Dalton Transactions, 2009, 2687-2694.

81. Campbell C.; Gordon S.; Smith C. L. *Derivative Thermoanalytical Techniques. Instrumentation and Applications to Thermogravimetry and Differential Thermal Anaylsis.* Analytical Chemistry, 1959. **31**, 1188-1191.

82. Cerenius Y.; Stahl K.; Svensson L. A.; Ursby T.; Oskarsson A.; Albertsson J.; Liljas A., *The crystallography beamline I711 at MAX II.* Journal of Synchrotron Radiation, 2000. **7**, 203-208.

83. Morgenroth W.; Overgaard J.; Clausen H. F.; Svendsen H.; Jorgensen M. R. V.; Larsen F. K.; Iversen B. B. *Helium cryostat synchrotron charge densities determined using a large CCD detector-the upgraded beamline D3 at DESY.* Journal of Applied Crystallography, 2008. **41**, 846-853.

84. Jensen T. R.; Nielsen T. K.; Filinchuk Y.; Jorgensen J.-E.; Cerenius Y.; Gray E. M.; Webb C. J., *Versatile in situ powder X-ray diffraction cells for solid-gas investigations.* Journal of Applied Crystallography, 2010. **43**, 1456-1463.

85. Ravel B.; Newville M. *ATHENA, ARTEMIS, HEPHAESTUS: data analysis for X-ray absorption spectroscopy using IFEFFIT.* Journal of Synchrotron Radiation, 2005. **12**, 537-541.

86. Newville M. *IFEFFIT : interactive XAFS analysis and FEFF fitting.* Journal of Synchrotron Radiation, 2001. **8**, 322-324.

87. Lutterotti L.; Matthies S.; Wenk H. R.; Schultz A. S.; Richardson J. W. *Combined texture and structure analysis of deformed limestone from time-of-flight neutron diffraction spectra.* Journal of Applied Physics, 1997. **81**, 594-600.

88. Friedrichs O.; Remhof A.; Borgschulte A.; Buchter F.; Orimo S. I.; Züttel A. *Breaking the passivation-the road to a solvent free borohydride synthesis.* Physical Chemistry Chemical Physics, 2010. **12**, 10919-10922.

89. AbdelDayem H. M.; Al-Omair M.A. *Phase Composition and Catalytic Activity of α-NiMoO$_4$ Reduced with Hydride Anion.* Industrial & Engineering Chemistry Research, 2008. **47**, 1011-1016.

90. Hwang S. J.; Bowman R. C.; Reiter J. W.; Rijssenbeek J.; Soloveichik G. L.; Zhao J. C.; Kabbour H.; Ahn C. C. *NMR confirmation for formation of $B_{12}H_{12}^{2-}$ complexes during hydrogen desorption from metal borohydrides.* Journal of Physical Chemistry C, 2008. **112**, 3164-3169.

91. Mean B. J.; Lee K. H.; Kang K. H.; Lee M.; Rhee J. S.; Cho B. K. *B-11 NMR study of calcium-hexaborides.* Physica B-Condensed Matter, 2005. **359**, 1204-1206.

92. Barin I. *Thermochemical Data of Pure Substances. Part I + II.* Weinheim:VCH VCH Verlagsgesellschaft, 1989. 1-816.

93. Bösenberg U.; Doppiu S.; Mosegaard L.; Barkhordarian G.; Eigen N.; Borgschulte A.; Jensen T. R.; Cerenius Y.; Gutfleisch O.; Klassen T.; Dornheim M.; Bormann R. *Hydrogen sorption properties of MgH_2-$LiBH_4$ composites.* Acta Materialia, 2007. **55**, 3951-3958.

94. Fan M.-Q.; Sun L.-X.; Zhang Y.; Xu F.; Zhang J.; Chu H.-L. *The catalytic effect of additive Nb_2O_5 on the reversible hydrogen storage performances of $LiBH_4$-MgH_2 composite.* International Journal of Hydrogen Energy, 2008. **33**, 74-80.

95. Ma L. P.; Kang X. D.; Dai H. B.; Liang Y.; Fang Z. Z.; Wang P. J.; Wang P.; Cheng H. M. *Superior catalytic effect of TiF_3 over $TiCl_3$ in improving the hydrogen sorption kinetics of MgH_2: Catalytic role of fluorine anion.* Acta Materialia, 2009. **57**, 2250-2258.

96. Zhao Y.; Feng Y.; Cheng C. H.; Zhou L.; Wu Y.; Machi T.; Fudamoto Y.; Koshizuka N.; Murakami M. *High critical current density of MgB_2 bulk superconductor doped with Ti and sintered at ambient pressure.* Applied Physics Letters, 2001. **79**, 1154-1156.

97. Jin S. A.; Shim J. H.; Ahn J. P.; Cho Y. W.; Yi K. W. *Improvement in hydrogen sorption kinetics of MgH_2 with Nb hydride catalyst.* Acta Materialia, 2007. **55**, 5073-5079.

98. Jin S. A.; Shim J. H.; Cho Y. W.; Yi K. W. *Dehydrogenation and hydrogenation characteristics of MgH_2 with transition metal fluorides.* Journal of Power Sources, 2007. **172**, 859-862.

99. Yavari A. R.; LeMoulec A.; de Castro F. R.; Deledda S.; Friedrichs O.; Botta W. J.; Vaughan G.; Klassen T.; Fernandez A.; Kvick A. *Improvement in H-sorption kinetics of MgH_2 powders by using Fe nanoparticles generated by reactive FeF_3 addition.* Scripta Materialia, 2005. **52**, 719-724.

100. Wang X. L.; Suda S. *Surface characteristics of fluorinated hydriding alloys.* Journal of Alloys and Compounds, 1995. **231**, 380-386.

101. Ivanov E.; Konstanchuk I.; Bokhonov B.; Boldyrev V. *Hydrogen interaction with mechanically alloyed magnesium–salt composite materials.* Journal of Alloys and Compounds, 2003. **359**, 320-325.

102. Buchter F.; Lodziana Z.; Remhof A.; Friedrichs O.; Borgschulte A.; Mauron P.; Züttel A.; Sheptyakov D.; Palatinus L.; Chlopek K.; Fichtner M.; Barkhordarian G.; Bormann R.; Hauback B. C. *Structure of the Orthorhombic gamma-Phase and Phase Transitions of $Ca(BD_4)_2$.* Journal of Physical Chemistry C, 2009. **113**, 17223-17230.

103. Deprez E.; Munoz-Marquez M. A.; Roldan M. A.; Prestipino C.; Palomares F. J.; Bonatto Minella C.; Bösenberg U.; Dornheim M.; Bormann R.; Fernandez A. *Oxidation State and Local Structure of Ti-Based Additives in the Reactive Hydride Composite 2LiBH$_4$ + MgH$_2$.* Journal of Physical Chemistry C, 2010. **114**, 3309-3317.

104. Brice J. F.; Courtois A.; Aubry J. *Preparation of a Hydrofluorinated Solid-Solution CaF$_{2-x}$H$_x$-Structural Determination Using X-ray and Neutron-Diffraction.* Journal of Solid State Chemistry, 1978. **24**, 381-387.

105. Takacs L.; McHenry J. S. *Temperature of the milling balls in shaker and planetary mills.* Journal of Materials Science, 2006. **41**, 5246-5249.

106. Borgschulte, A., et al., *Experimental evidence of librational vibrations determining the stability of calcium borohydride.* Physical Review B, 2011. **83**(2).

107. Friedrichs O.; Aguey-Zinsou F.; Fernández J. R. A.; Sánchez-López J. C.; Justo A.; Klassen T.; Bormann R.; Fernández A. *MgH$_2$ with Nb$_2$O$_5$ as additive, for hydrogen storage: Chemical, structural and kinetic behavior with heating.* Acta Materialia, 2006. **54**, 105-110.

108. Friedrichs O.; Martínez-Martínez D.; Guilera G.; Sánchez López J. C.; Fernández A. *In Situ Energy-Dispersive XAS and XRD Study of the Superior Hydrogen Storage System MgH$_2$/Nb$_2$O$_5$.* The Journal of Physical Chemistry C, 2007. **111**, 10700-10706.

109. Friedrichs O.; Sánchez-López J. C.; López-Cartes C.; Klassen T.; Bormann R.; Fernández A. *Nb$_2$O$_5$ "Pathway Effect" on Hydrogen Sorption in Mg.* The Journal of Physical Chemistry B, 2006. **110**, 7845-7850.

110. Deprez E.; Munoz-Marquez M. A.; de Haro M. C. J.; Palomares F. J.; Soria F.; Dornheim M.; Bormann R.; Fernandez A., *Combined x-ray photoelectron spectroscopy and scanning electron microscopy studies of the LiBH$_4$-MgH$_2$ reactive hydride composite with and without a Ti-based additive.* Journal of Applied Physics, 2011. **109**.

111. Bösenberg U.; Vainio U.; Pranzas P. K.; Bellosta von Colbe J. M.; Goerigk G.; Welter E.; Dornheim M.; Schreyer A.; Bormann R. *On the chemical state and distribution of Zr- and V-based additives in reactive hydride composites.* Nanotechnology, 2009. **20**, 204003.

112. Bösenberg U.; Kim J. W.; Gosslar D.; Eigen N.; Jensen T. R.; Bellosta von Colbe J. M.; Zhou Y.; Dahms M.; Kim D. H.; Gunther R.; Cho Y. W.; Oh K. H.; Klassen T.; Bormann R.; Dornheim M. *Role of additives in LiBH$_4$-MgH$_2$ reactive hydride composites for sorption kinetics.* Acta Materialia, 2010. **58**, 3381-3389.

113. Ngene P.; Verkuijlen M. H. W.; Zheng Q.; Kragten J.; van Bentum P. J. M.; Bitter J. H.; de Jongh P. E. *The role of Ni in increasing the reversibility of the hydrogen release from nanoconfined LiBH$_4$.* Faraday Discussions, 2011.

114. Buslaev Y. A.; Dyer D. S.; Ragsdale R. O. *Hydrolysis of titanium tetrafluoride.* Inorganic Chemistry, 1967. **6**, 2208-2212.

115. Hermanek S., *Boron-11 NMR spectra of boranes, main-group heteroboranes, and substituted derivatives. Factors influencing chemical shifts of skeletal atoms.* Chemical Reviews, 1992. **92**, 325-362.

116. http://www.chemistry.sdsu.edu/research/BNMR/#summary.

117. Gingl F.; Bonhomme F.; Yvon K.; Fischer P. *Tetracalcium Trimagnesium Tetradecahydride, $Ca_4Mg_3H_{14}$ - The 1st Ternary Alkaline-Earth Hydride.* Journal of Alloys and Compounds, 1992. **185**, 273-278.

118. Kim Y.; Reed D.; Lee Y. S.; Shim J. H.; Han H. N.; Book D.; Cho Y. W. *Hydrogenation reaction of CaH_2-CaB_6-Mg mixture.* Journal of Alloys and Compounds, 2010. **492**, 597-600.

119. Reiter J. W.; Zan J. A.; Hwang S.-J. *Development and Evaluation of Advanced Hydride Systems for Reversible Hydrogen Storage.* FY 2010 Annual Progress Report, 2010.

120. Pistidda C.; Garroni S.; Dolci F.; Bardají E. G.; Khandelwal A.; Nolis P.; Dornheim M.; Gosalawit R.; Jensen T.; Cerenius Y.; Suriñach S.; Baró M. D.; Lohstroh W.; Fichtner M. *Synthesis of amorphous $Mg(BH_4)_2$ from MgB_2 and H_2 at room temperature.* Journal of Alloys and Compounds, 2010. **508**, 212-215.

121. Bonatto Minella C.; Garroni S.; Pistidda C.; Gosalawit-Utke R.; Barkhordarian G.; Rongeat C.; Lindemann I.; Gutfleisch O.; Jensen T. R.; Cerenius Y.; Christensen J.; Baro M. D.; Bormann R.; Klassen T.; Dornheim M. *Effect of Transition Metal Fluorides on the Sorption Properties and Reversible Formation of $Ca(BH_4)_2$.* Journal of Physical Chemistry C, 2011. **115**, 2497-2504.

122. Mauron P.; Buchter F.; Friedrichs O.; Remhof A.; Bielmann M.; Zwicky C. N.; Züttel, A. *Stability and Reversibility of $LiBH_4$.* The Journal of Physical Chemistry B, 2007. **112**, 906-910.

123. Kim Y.; Hwang S.-J.; Shim J.-H.; Lee Y.-S.; Han H. N.; Cho Y. W. *Investigation of the Dehydrogenation Reaction Pathway of $Ca(BH_4)_2$ and Reversibility of Intermediate Phases.* The Journal of Physical Chemistry C, 2012. **116**, 4330-4334.

124. Konings R. *The composition of niobium pentafluoride vapor.* Structural Chemistry, 1994. **5**, 9-13.

125. Zhang M. X.; Kelly P. M. *Edge-to-edge matching model for predicting orientation relationships and habit planes-the improvements.* Scripta Materialia, 2005. **52**, 963-968.

126. Schiavo B.; Girella A.; Agresti F.; Capurso G.; Milanese C. *Ball-milling and AlB_2 addition effects on the hydrogen sorption properties of the CaH_2 + MgB_2 system.* Journal of Alloys and Compounds, 2011. **509, Supplement 2**, S714-S718.

127. Stavila V.; Her J. H.; Zhou W.; Hwang S. J.; Kim C.; Ottley L. A. M.; Udovic T. J. *Probing the structure, stability and hydrogen storage properties of calcium dodecahydro-closo-dodecaborate.* Journal of Solid State Chemistry, 2010. **183**, 1133-1140.

128. Bosenberg U.; Ravnsbaek D. B.; Hagemann H.; D'Anna V.; Bonatto Minella C.; Pistidda C.; van Beek W.; Jensen T. R.; Bormann R.; Dornheim M. *Pressure and Temperature Influence on the Desorption Pathway of the $LiBH_4$−MgH_2 Composite System.* The Journal of Physical Chemistry C, 2010. **114**, 15212-15217.

129. Newhouse R. J.; Stavila V.; Hwang S.-J.; Klebanoff L. E.; Zhang J. Z. *Reversibility and Improved Hydrogen Release of Magnesium Borohydride*. The Journal of Physical Chemistry C, 2010. **114**, 5224-5232.

130. Bonatto Minella C.; Garroni S.; Olid D.; Teixidor F.; Pistidda C.; Lindemann I.; Gutfleisch O.; Baró M. D.; Bormann R.; Klassen T.; Dornheim M. *Experimental Evidence of $Ca[B_{12}H_{12}]$ Formation During Decomposition of a $Ca(BH_4)_2$ + MgH_2 Based Reactive Hydride Composite*. The Journal of Physical Chemistry C, 2011. **115**, 18010-18014.

131. Ngene P.; van den Berg R.; Verkuijlen M. H. W.; de Jong K. P.; de Jongh P. E. *Reversibility of the hydrogen desorption from $NaBH_4$ by confinement in nanoporous carbon*. Energy & Environmental Science, 2011. **4**, 4108

7 Acknowledgments

I am very grateful to Dr. Martin Dornheim and Prof. Rüdiger Bormann for the supervision of this work and for giving me the chance to join their research group in the field of hydrogen storage. Working in a European network which was the COSY project represented for me a great professional chance.

I would also like to thank Prof. Thomas Klassen for his constant encouragement and his extremely positive spirit which has inspired me.

I am very grateful to Prof. Oliver Gutfleisch for offering me the opportunity to work in his laboratories at the Institute for Metallic Materials at the IFW in Dresden.

I would like to thank Prof. Duncan H. Gregory from the University of Glasgow. Duncan has a courteousness which I saw rarely in life. I am glad I had the privilege to meet him.

In addition, I wish to thank all the people and/or Institutes listed here below:

• All the colleagues of the Department of Nanotechnology of the Institute of Materials Research of the Helmholtz-Zentrum Geesthacht.

• The staff of the Research Department of Functional Magnetic Materials and Hydrides at the Institute for Metallic Materials, IFW Dresden. Prof. Ludwig Schultz, Inge Lindeman, Christian Geipel, Dr. Roger Domenech-Ferrer, Dr. Carine Rongeat. In particular, I express my gratitude to Monika Herrich and Bernhard Gebel for the support in the laboratory.

• Prof. Asunción Fernandez and Dr. Emilie Deprez for giving me the chance to spend two weeks at their Institute (ICMS) in Seville. Emilie was extremely kind in helping me to analysis the XAS data collected at the synchrotron.

• Dr. Torben R. Jensen for giving me the possibility to analyse the materials by *in situ*-XRD at the beamline I711 in Lund. I am thankful to the beamline staff as well. Above all, I would like to mention Dorthe Haase for her competence and friendliness.

• Dr. Chiara Milanese from the University of Pavia. She is always extremely supportive.

• Prof. Dolors Baró and his group for the TEM and NMR data. Thanks to Dr. Eva Pellicer and Dr. Emma Rossinyol (UAB, Barcelona) for collecting the TEM pictures. I know how complex it was to analyse those samples. Thanks to Dr. Pau Nolis and the Servei de Ressonancia Magnetica Nuclear RMN at UAB for their technical assistance.

I desire to thank Sebastiano and Claudio for the pleasant moments spent together.

I am grateful to Elsa who has been very supportive in the final part of the writing process. Her advices and friendship are precious to me.

I would like to thank my family who has always being supportive during these years.

Finally, I am grateful to Anja. This work is dedicated to her. She is the only one able to make me a better person.

I want morebooks!

Buy your books fast and straightforward online - at one of the world's fastest growing online book stores! Environmentally sound due to Print-on-Demand technologies.

Buy your books online at
www.get-morebooks.com

Kaufen Sie Ihre Bücher schnell und unkompliziert online – auf einer der am schnellsten wachsenden Buchhandelsplattformen weltweit! Dank Print-On-Demand umwelt- und ressourcenschonend produziert.

Bücher schneller online kaufen
www.morebooks.de

VDM Verlagsservicegesellschaft mbH
Heinrich-Böcking-Str. 6-8
D - 66121 Saarbrücken   Telefax: +49 681 93 81 567-9

info@vdm-vsg.de
www.vdm-vsg.de

Printed by Books on Demand GmbH, Norderstedt / Germany